KINETICS OF ENZYMATIC SYNTHESIS

Edited by **Lakshmanan Rajendran**
and **Carlos Fernandez**

Kinetics of Enzymatic Synthesis
http://dx.doi.org/10.5772/intechopen.76270
Edited by Lakshmanan Rajendran and Carlos Fernandez

Contributors

Christoph Held, Gabriele Sadowski, Anton Wangler, Mark Jonathan Bunse, Alma Hortensia Serafin Muñoz, Berenice Noriega Luna, Julio Cesar Leal Vaca, Carlos Eduardo Molina-Guerrero, Aurelio Alvarez-Vargas, Maria Isabel Estrada, Luis A. Cira-Chávez, Joseph Guevara Luna, Marisela Yadira Soto Padilla, Brenda Román Ponce, María Vásquez-Murrieta, Syamsul Rizal Abd Shukor, Nurhazwani Yusoff Azudin, Elena Grosu, Ionela Ungureanu, Kalim Belhacene, Renato Froidevaux, Alexandra Blaga, Pascal Dhulster, Remi Przybylski

First published in London, United Kingdom, 2019 by IntechOpen
IntechOpen is the global imprint of INTECHOPEN LIMITED, registered in England and Wales, registration number: 11086078, The Shard, 25th floor, 32 London Bridge Street
London, SE19SG – United Kingdom
Printed in Croatia

British Library Cataloguing-in-Publication Data
A catalogue record for this book is available from the British Library

Additional hard copies can be obtained from orders@intechopen.com

Kinetics of Enzymatic Synthesis, Edited by Lakshmanan Rajendran and Carlos Fernandez
p. cm.
Print ISBN 978-1-78985-029-1
Online ISBN 978-1-78985-030-7

For EU product safety concerns:
IN TECH d.o.o., Prolaz Marije Krucifikse Kozulić 3, 51000 Rijeka, Croatia,
info@intechopen.com or visit our website at intechopen.com.

Meet the editors

Dr. Lakshmanan Rajendran is a professor at the Department of Mathematics, Academy of Maritime Education and Training (deemed to be a university), Chennai, India. He has published 110 papers in international/SCI journals and 100 papers in national journals. He has also written four books. He serves as a reviewer in many international journals. He completed six research projects from various funding agencies in India. More than 35 students completed their PhD degrees under his guidance. He visited Germany and Poland under INSA fellowships. His research interest is mathematical modeling and nonlinear reaction diffusion processes in biosensors, biofuel cells, and bioreactors.

Dr. Carlos Fernandez is a lecturer in analytical chemistry at Robert Gordon University. He has published over 50 peer-reviewed articles. His research interests focus on the use of voltammetric techniques in analytical chemistry as an electrochemical sensor to detect the following analytes: heavy metals for environmental applications, drugs of abuse, amino acids, and pharmaceutical drugs with and without graphene-based compounds. Furthermore, his research emphasizes the utilization of graphene-based compounds and nanocomposites for energy storage devices. He is also interested in the investigation of corrosion processes using electrochemical techniques.

Contents

Preface

Kinetics of Enzymatic Synthesis contains five chapters focused on advanced methods used in the research area of kinetics of enzymatic synthesis. This book was written to explain the fundamental as well as advanced application of enzyme kinetics and provides an overview on some of the recent developments in the kinetics of enzymatic synthesis. Particular emphasis is given to both theoretical and experimental aspects of the synthesis of enzymes. The five chapters of the book serve as an important reference for scientists and academics working in the research areas mentioned above, especially in the aspects of kinetics of enzymatic synthesis.

The first section "Enzyme Kinetics" contains two chapters with a wide range of topics. In recent years, halophilic microorganisms have been explored for their biotechnological potential in different fields. The first chapter focuses on the kinetics of halophilic enzymes, which have several biotechnological applications. The classical thermodynamic approach towards analyzing the influence of co-solvents on the Michaelis constants of enzyme-catalyzed reactions is demonstrated in the second chapter.

The second section "Enzymatic Synthesis" covers three chapters. Production of isoamyl acetate using the enzymatic synthesis method between acetic anhydride and isoamyl alcohol by having the enzyme *Candida antarctica* Lipase B as a catalyst in a solvent-free system is discussed in the third chapter. In this study the optimization process and mathematical modeling of the kinetic reaction and yield of isoamyl acetate are included. The integrated scheme with the use of the filtrate from the pretreatment of the CS and the growth conditions of *Pleurotuscystidiosus* is studied in the fourth chapter. The last chapter of this section provides the conditions of the key parameters in microfluidic systems (residence times, flow rates, concentrations) applied for a sequential process from liquid/liquid extraction of LVV-h7.

I am thankful to Shri J. Ramachandran, Chancellor, and Col. Dr. G. Thiruvasagam, Vice-Chancellor, Academy of Maritime Education and Training, deemed to be a university, for their constant encouragement. Special thanks go to co-editor Dr. Carlos Fernandez, School of Pharmacy and Life Sciences, Robert Gordon University, Aberdeen (UK), for valuable suggestions.

I would like to thank Author Service Manager Ms. Marijana Francetic for her cooperation throughout the process of the publication of this book.

Dr. Lakshmanan Rajendran
Academy of Maritime Education and Training (AMET)
Deemed to be University
Chennai, Tamilnadu, India

Dr. Carlos Fernandez
Robert Gordon University
Aberdeen, UK

Section 1

Enzyme Kinetics

Chapter 1

Kinetics of Halophilic Enzymes

Luis Alberto Cira-Chávez, Joseph Guevara-Luna,
Marisela Yadira Soto-Padilla, Brenda Román-Ponce,
María Soledad Vásquez-Murrieta and
María Isabel Estrada-Alvarado

Additional information is available at the end of the chapter

http://dx.doi.org/10.5772/intechopen.81100

Abstract

Hypersaline environments are those with salt concentrations 9–10 times higher (30–35% of NaCl) than sea water (3.5% of NaCl). At high concentrations of soluble salts, cytoplasm—mainly of bacteria and archaea—is exposed to high ionic strength and achieves osmotic equilibrium by maintaining a cytoplasmic salt concentration similar to that of the surrounding media. Halophilic enzymes are extremozymes produced by halophilic microorganisms; they have similar characteristics to regular enzymes but different properties, mainly structural. Among these properties is a high requirement of salt for biological functions. Furthermore, the discovery of enzymes capable of degrading biopolymers offer a new perspective in the treatment of residues from oil deposits, under typically high conditions of salt and temperature, while giving valuable information on heterotrophic processes in saline environments.

Keywords: halotolerants, halophiles, *salt-in*, synthesis intracellular compounds, extremonzymes

1. Introduction

Extreme environments involve a wide range of extreme conditions (pH, temperature, pressure, light intensity, oxygen, nutrient conditions, heavy metals, and salinity). Hypersaline environments are those with salt concentrations 9–10 times higher (30–35% of NaCl) than sea water (3.5% of NaCl). These sites are widely distributed around the world and can harbor microorganisms from three different life domains (archaea, bacteria, and eukaryota); together, these microorganisms are known as halophiles, which survive or even thrive in saline environments [1].

2. Classification of halophiles and halophile environments

Nowadays, several classifications of halophiles have been suggested; the classification proposed by Ollivier et al. [2] considers those microorganisms capable of growing in salt concentrations ≥150 g L^{-1} (15% w/v, 2.5 M) as halophiles. Another classification considers the optimum growth salinity as follows: mild halophiles (1–6%, w/v NaCl), moderate halophiles (7–15%) and extreme halophiles (15–30%) [3]. On the other hand, Ventosa and Arahal [4] defines halophiles as organisms that have an optimal growth above 3% salt concentration; if the optimal growth occurs between 3 and 15% salt, they are regarded as moderate halophiles; and when it occurs above 15% and up to halite saturation (34%), they are regarded as extreme halophiles. In addition, DasSarma and DasSarma [5] described halophiles as those organisms that thrive from sea salinity (~0.6 M) up to saturation salinity (>5 M NaCl). However, the most complete and widely used classification scheme was proposed by Kushner and Kamekura [6], in which halophilic microorganisms are separated into six groups based on their salt requirement and tolerance (**Table 1**): non-halophiles are those that have optimal growth in culture media containing less than 0.2 M NaCl; slight halophiles (marine bacteria) grow best in media with 0.2–0.5 M NaCl; moderate halophiles grow best with 0.5–2.5 M NaCl; borderline extreme halophile that growth best at 2.5–4.0 M; extreme halophiles show optimal growth in culture media containing NaCl concentration between 4 and 5.9 M; and finally halo-tolerant microorganisms, which are non-halophiles that can tolerate high salt concentrations but do not require salt to survive; any microorganism viable at 2.5 M of NaCl is considered extremely halotolerant. Archaea and bacteria are the most widely distributed organisms in hypersaline environments [7], especially in those in which salinities exceed 1.5 M (about 10%). In recent years, halophilic organisms are mainly isolated from saline environments, such as salt lakes, marine solar salterns, saline soils, and marine sediments (see **Table 1**). However, halophile bacteria have also been isolated from some non-common places, for example, textile effluents, halophytes, mine tailings as well as processed foods (**Table 1**).

2.1. Hypersaline environments

Hypersaline environments are extreme habitats with limited microbial diversity as result of high salt concentrations and other environmental factors. Nowadays, most environmental studies have been carried out on aquatic habitats, such as saline lakes and solar salterns used for the production of salt for commercial purposes [8]. Nevertheless, halophilic bacteria can be found in other habitats including saline soils, salted foods and other products, hides, and deep-sea brine pools [7, 9–11]. Depending on whether they originated or not from seawater, hypersaline environments are classified as thalassohaline and athalassohaline, respectively.

2.1.1. Thalassohaline environments

The thalassohaline environments are saline environments of marine origin, which contain the following ions: Cl^{-1}, Na^{+}, Mg^{2+}, SO_4^{2-}, K^{+}, Ca^{2+}, Br^{-}, HCO_3^{-}, and F^{-} [4]. Some examples of thalassohaline as explained as follows.

Category	Salt tolerance (M)	Example	Isolation site	References
Non-halophile	<0.2	*Vibrio palustris* EAdo9^T and *Vibrio spartinae* SMJ221^T	Salt-marsh plants	[106]
Slight halophile	0.2–0.5	*Paracoccus* sp. GSM2	Textile mill effluent	[107]
		Bacillus sp. NY6	Saline wastewater	[108]
		Zunongwangia endophytica CPA58^T	Tissues of the halophyte *Halimione portulacoides*	[109]
Moderate halophile	0.5–2.5	*Salinispora arenicola* CNH-643^T and *Salinispora tropica* CNB-440^T	Marine sediments	[110]
		Salinispora pacifica CNR-114^T	Marine sediments	[111]
		Martelella endophytica YC6887^T	Root of *Rosa rugosa*	[112]
		Streptomyces halophyticola KLBMP 1284^T	Stems of *Tamarix chinensis*	[113]
		Labrenzia suaedae YC6927^T	Root of *Sauceda maritime*	[114]
		Kocuria arsenatis CM1E1^T	*Prosopis laevigata*	[115]
		Proteus sp. NA6	Textile effluent drain	[116]
		Candidatus *Desulfonatronobulbus propionicus*	Hypersaline soda lakes	[117]
		Novosphingobium pokkalii L3E4^T	Rhizosphere of saline-tolerant pokkali rice	[118]
		Marinobacter aquaticus M6-53^T	Marine saltern located in Huelva, Spain	[119]
		Agrobacterium salinitolerans YIC 5082^T	Root nodules of *Sesbania cannabina* grown in a high-salt and alkaline environment	[120]
		Salinicola tamaricis F01^T	Leaves of *Tamarix chinensis*	[121]
		Salinirubellus salinus ZS-35-S2^T	Marine solar saltern	[122]
Borderline extreme halophile	2.5–4.0	*Aliifodinibius halophilus* 2W32^T	Marine solar saltern	[123]
		Desulfosalsimonas propionicica PropA^T	Hypersaline sediment of the Great Salt Lake	[124]
		Salinibacter iranicus CB7^T and *Salinibacter luteus* DGO^T	Aran-Bidgol salt lake, Iran	[125]
		Halanaerobium sehlinense 1Sehel^T	Sediments of the hypersaline lake Sehline Sebkha	[126]
		Sporohalobacter salinus CEJFT1B^T	Under the salt crust of El-Jerid hypersaline lake in southern Tunisia	[127]
		Lentibacillus kimchii K9^T	Korean fermented food (kimchi)	[128]
		Marinobacter salexigens HJR7^T	Marine sediment	[129]
		Gracilimonas halophila WDS2C40^T	Marine solar saltern	[130]
		Salinifilum proteinilyticum Miq-12^T	Wetland in Iran	[131]
		Natronospira proteinivora Bsker1^T	Marine solar saltern	[132]

Category	Salt tolerance (M)	Example	Isolation site	References
Extreme halophile	4–5.9	*Salinibacter ruber* M31^T	Saltern crystallizer ponds in Alicante and Mallorca, Spain	[12]
		Limimonas halophila IA16^T	Mud of the hypersaline Lake Aran-Bidgol, Iran	[133]
		Desulfonatronobacter acetoxydans APT3	Hypersaline soda lake	[47]
Halo-tolerant	A non-halophile that tolerant salt; if it is viable 2.5 M, in is considered extremely halotolerant	*Brevibacterium salitolerans* TRM 415^T	Sediment from a salt lake	[134]
		Kineococcus endophytica KLMMP 1274^T	Halophytic plant (*Limonium sinense*)	[135]
		Anditalea andensis ANESC-S^T	Alkali-saline soil	[136]
		Brevibacterium jeotgali SJ5-8^T	Traditional Korean fermented seafood	[137]
		Salimicrobium sp. LY19	Saline soil	[138]
		Brevibacterium metallicus NM3E2^T	Edge of mine tailings	[139]
		Bacillus subtilis BLK-1.5	Salt mines	[76]
		Halomonas nigrificans MB G8645^T	Acid curd cheese called Quargel	[140]

Table 1. Classification of bacteria based on their salinity tolerance according with criteria proposed by Kushner and Kamekura [6].

2.1.1.1. Solar salterns

These sites have a similar composition to seawater and they are used for salt production by evaporation. They generally consist of several ponds interconnected to form the so-called multipond system. Seawater is pumped or allowed to flow into the first ponds, and as a consequence of solar evaporation, the concentration of salts increases slightly and the water is moved to the next ponds, where it will concentrate further. Finally, in the last pond (called crystallizer), common salt is precipitated [4]. Many studies have focused on the isolation of bacteria harbored in hypersaline environments, identifying the following major groups: Bacteroidetes [12], Firmicutes [13–15], ϒ-Proteobacteria [13, 16–17], and ϒ-Proteobacteria being the most abundant.

2.1.1.2. Soils

Saline soils are those with an electrical conductivity (EC) higher than 4 dS mL^{-1}, approximately 40 mM NaCl [18]. Nowadays, salinized areas are increasing at rate of 10% annually for various reasons, including low precipitation, high surface evaporation, weathering of native rocks, irrigation with saline water, and poor cultural species practices [19]. Several studies on hypersaline soils mainly isolated moderate halophiles and non-halophilic bacteria affiliated with different genera of the following taxonomic groups: Firmicutes [20–22], actinobacteria [23, 24], and proteobacteria [25, 26].

2.1.1.3. *Great Salt Lake*

In 1957, a rock-filled railroad causeway was completed across the lake, dividing it into a northern and a southern basin. The northern arm presented high salinity (33%), while the southern arm separated by a semipermeable rock causeway contains a moderate concentration of salt (12%) [27]. A number of studies on the isolation of bacteria from the sediment from Great Salt Lake have been carried out [28, 29].

2.1.2. *Athalassohaline environments*

These are environments that do not have a marine origin and their ionic proportions are quite different from that of the dissolved salts in seawater [30]. They reflect the composition of the surrounding geology, topography, and climate conditions, often particularly influenced by the dissolution of mineral deposits [31].

2.1.2.1. *Dead Sea*

The Dead Sea, which is actually an inland lake, is famous for being so saline that people can float with ease on its surface. The site is composed mainly of divalent ions like Mg^{2+} [32]. The Dead Sea is a hypersaline lake with 34% salinity, and its name is due to the lack of any living macroscopic creatures. The lake consists of a deeper northern basin and a shallow southern basin, which has been recently dried up and used for commercial mineral production [33]. The water level is dependent on the balance between amount of freshwater inflow and evaporation [32]. The Jordan River is the main source of freshwater inflow, in addition to several water springs and the complex system of underwater springs, which has been recently discovered [34]. Metagenomic studies demonstrated the presence of Halobacterium-like sequences and Mg^{2+} transport-related proteins, suggesting a potential adaptation to the high magnesium concentration by Dead Sea halophiles [35]. In addition, metagenomic sequence analysis and amino acid profiling also demonstrated the presence of halophiles never previously isolated or sequenced in the Dead Sea [35, 36]. It was recently discovered that the Dead Sea harbors some bacteria with biotechnological properties, such as *Bacillus persicus* 24-DSM, which showed antimicrobial activity [33].

2.1.2.2. *Soda lakes*

Athalassohaline alkaline salt lakes (or soda lakes), rich in NaCl, $NaHCO_3$, and Na_2CO_3, are usually formed by dissolution of rocks that are low in magnesium and calcium, which would otherwise cause carbonate to precipitate [37, 38]. The most studied lakes are those in the East African Rift Valley, continental Russia, and the USA. In addition to being able to tolerate high pH values and elevated salinities, microbes inhabiting soda lakes have to cope with low availability of NH_4^+, caused by weak dissociation of ammonia at high pH. An accumulation of stressful but volatile NH_3 may occur in enclosed alkaline-saline systems such as sea ice or locally in soda lakes [32]. Several studies about bacteria isolation from soda lakes from different continents are available [39–47].

Halophilic microorganisms have several biotechnological applications, such as β-carotene production of fermented foods. In recent years, uses of halophilic microorganisms have significantly increased. Many enzymes, stabilizers, and valuable compounds from halophiles may present advantages for the development of biotechnological production processes.

2.2. Biology and adaptation of halophilic bacteria

The first chemical stress encountered during the evolution of life on earth may have been salt stress. Thus, from the beginning, organisms must have evolved strategies and effective mechanisms for the stabilization of protoplasmic structures and ion regulation [48]. At high concentrations of soluble salts, cytoplasm—mainly of bacteria and archaea—is exposed to high ionic strength and achieves osmotic equilibrium by maintaining a cytoplasmic salt concentration similar to that of the surrounding media. This can affect microbes via two primary mechanisms: osmotic effect and specific ion effects. Soluble salts increase the osmotic potential (more negative) of the soil water, drawing water out of cells which may kill microbes and roots through plasmolysis [49, 50].

To thrive in the hypersaline environment, halophiles have two main adaptation mechanisms to prevent NaCl from diffusing into the cells. The first mechanism is accumulation of inorganic ions (mainly KCl) for balancing osmotic pressure. This mechanism is mainly utilized by aerobic and extremely halophilic archaea and some anaerobic halophilic bacteria [32, 49, 51]. In contrast, most halophilic bacteria accumulate water soluble organic compounds of low molecular weight, which are referred to as compatible solutes or osmolytes, to maintain low intracellular salt concentration [52–54].

2.2.1. Salt-in mechanisms

As mentioned above, microorganisms that grow optimally in the presence of extremely high salinities (up to 5 M NaCl), accumulate intracellular potassium and chloride ions in concentrations higher than the external NaCl concentration to maintain a turgor pressure. This so-called "salt-in" strategy is observed in *Halobacteriales* (archaea) and *Halanaerobiales* (anaerobic halophilic bacteria) [55, 56]. The mechanism ("salt-in" to balance "salt-out") requires far-reaching adaptations of the entire intracellular machinery, as all enzymes and functions in the cytoplasm have to be functional in the presence of molar concentrations of KCl [57]. A characteristic feature of halophilic proteins from microorganisms that accumulate KCl for osmotic balance is their highly acidic nature, with a great excess of acidic amino acids (glutamate and aspartate) over basic amino acids (lysine and arginine). Such proteins are highly negatively charged compared to their non-halophilic equivalents. In addition, halophilic proteins generally have a low content of hydrophobic amino acids [58, 59].

2.2.2. *Synthesis of intracellular compounds*

As explained above, microorganisms have the ability to adapt to or tolerate stress caused by salinity by accumulating osmolytes, also known as compatible solutes. The compatible solute strategy is broadly known in domain archaea, bacteria, as well as eukarya. Organisms accumulate organic solutes by uptake from the environment or *de novo* synthesis of organic compounds, such as sugars and polyols, amino acids and their derivatives, and other compatible solutes for protection against salinity stress [60–62].

Organic solutes act as stabilizers for biological structures and allow the cells to adapt not only to salts but also to heat, desiccation, cold, or even freezing conditions [63]. Many halophilic bacteria accumulate ectoine or hydroxyectoine as the predominant compatible solutes. Other intracellular compatible solutes include amino acids, glycine betaine and other compounds accumulated in small amounts [54].

Mei et al. [64] describe the physiology of a *Natrinema* sp. strain J7–2, an extremely halophilic archaea isolated from a salt mine in China, under salt stress conditions (15, 25, and 30% NaCl). This strain showed the highest growth rate at 25–30% of NaCl, while at 15% cells were more fragile. Furthermore, the glycerolipid and amino acidic metabolism showed a significant difference in cellular transcripts levels, perhaps playing a role in membrane production/alteration or in accumulation of specific amino acids (glutamate family—Glu, Arg and Pro; aspartate family—Asp; and aromatic amino acids—Phe and Trp), especially Glu and Asp as carbon substrates and energy resources or compatible solutes.

The most common inorganic solutes used as osmolytes by salinity tolerant microbes are potassium cations, while proline and glycine betaine are the main organic osmolytes [65]. However, the synthesis of these compounds requires high amounts of energy [50, 66]. Given these high energetic requirements, there are few reports of halophilic microorganisms that can produce compatible solutes to mitigate the stress by variable concentrations of salts. The capacity of two halophilic strains is noteworthy: *Planococcus* sp. VITP21 and *Bacillus* sp. VITP4, which are capable of *de novo* synthesis of two rarely occurring diamino acids, Nε-acetyl α-lysine and Nδ-acetyl ornithine, respectively; besides the well-known ectoine and proline [67] as simple diamino acidic molecules to tolerate salt stress.

2.3. Production of extremozymes

Halophilic enzymes are extremozymes produced by halophilic microorganisms; they have similar characteristics to regular enzymes but different properties, mainly structural. Among these properties is a high requirement of salt for biological functions. In recent years, different studies have focused on the detection of halophiles in saline environments in order to isolate and characterize new enzymatic activities. This resulted in several halophile hydrolases being described, including amylases, lipases, and proteases. Furthermore, the discovery of enzymes capable of degrading biopolymers offer a new perspective in the treatment of residues from oil deposits, under typically high conditions of salt and temperature, while giving valuable information on heterotrophic processes in saline environments.

2.3.1. Extremozymes-producing halophiles

Nowadays, investigation on the production of extremozymes from different bacterial genus and halophilic archaea has intensified. This interest is due to their capacity to efficiently catalyze a process and show optimal activities at different salt concentrations. Halophiles are the most probable source of extremozymes, since they are also capable of tolerating alkaline pH and high temperatures, as reported by several authors [68–79].

Most of the evaluation studies on the enzymatic capacities of halophiles begin with the isolation of these microorganisms from environments considered extreme due to specific characteristics such as high salt concentrations, high pH values, and extreme temperature conditions. Sánchez-Porro et al. [80] report the isolation of moderately halophile strains from water and salterns in different areas of southern Spain: Almería (Cabo de Gata), Cádiz (San Vicente and San Fernando), and Huelva (Isla Bacuta, Río Tinto and Isla Cristina). Isolates have been identified as members of the genera *Salinivibrio, Bacillus, Salibacillus, Halomonas, Chromohalobacter, Salinicoccus*, and *Marinococcus* and they showed amylase, protease, lipase, and DNase activities.

In 2007, Vidyasagar et al. [68] isolated the extreme halophile *Chromohalobacter* sp. from solar lanterns, and subsequently produced and partially purified a halo-thermophile protease extracellular enzyme. *Chromohalobacter* sp. required a 4 M concentration of NaCl for optimal growth and protease secretion, and no growth was observed under 1 M NaCl. The initial pH of the medium for growth and enzyme production was in the interval of 7.0–8.0, with an optimum value of 7.2. Halophile *Salinivibrio* sp. isolated from Bakhtegan Lake in southern Iran also produced an extracellular protease [81].

Rohban et al. [82] studied extremophiles in Howz Soltan, a hypersaline lake located in central Iran. The organisms successfully isolated produced a wide variety of extracellular enzymes, where 84.4% had lipase activity, 76.6% amylase, 43.2% protease, 41.1% inulinase, 39.8% xylanase, 29.4% cellulase, 14.2% DNase, and 12.1% pectinase. Halophile strains were identified as members of the following genera: *Salicola, Halovibrio, Halomonas, Oceanobacillus, Thalassobacillus, Halobacillus, Virgibacillus, Gracilibacillus, Salinicoccus*, and *Piscibacillus*. Most of the lipase and DNase producers belonged to the *Gracilibacillus* and *Halomonas* genera, respectively, while most of the organisms capable of producing hydrolytic enzymes (amylase, protease, cellulase, and inulinase) were part of Gram-positive genera, such as *Gracilibacillus, Thalassobacillus, Virgibacillus*, and *Halobacillus*.

In 2011, Perez et al. [83] reported the isolation and purification of a lipase obtained from the *Marinobacter lipolyticus* SM19 halophile, isolated from a saline habitat in southern Spain. The properties of this enzyme are of great potential for the food industry. Li and Yu [84] isolated the halophile strain LY9, which has amylolytic properties, from soil samples obtained in Yuncheng, China. The strain LY9 was identified as a member of the halobacillus genus and it was discovered that the production of amylase secreted for this strain depended on the salinity of the growth medium. The maximum production of amylase was observed in the presence of 10% KCl or 10% NaCl. Maltose was the main product from hydrolysis of soluble starch, pointing out to β-amylase activity.

In agreement with previous studies from 2003, 2007, and 2009, Shahbazi and Karbalaei-Heidari [85] reported the capacity of *Salinivibrio* sp. to produce extracellular low molecular weight proteases, and Jayachandra et al. [86] reported the isolation and identification of extracellular activity of hydrolytic enzymes from bacteria in the *Salinicoccus* sp. genus. Strain JAS4 was isolated from the Arabal soil in the west coast of Karnataka, India. These bacteria showed great potential to produce extracellular enzymes such as amylase, protease, inulinase, and gelatinase. Also in 2012, Kumar et al. [70] isolated halophiles from different saline environments in India, by means of morphological, biochemical, and 16S rRNA analyses. The authors identified the genera *Marinobacter, Virgibacillus, Halobacillus, Geomicrobium, Chromohalobacter, Oceanobacillus, Bacillus, Halomonas*, and *Staphylococcus* as having hydrolase activities of industrial relevance, pointing out the presence of amylases, lipases, and proteases. A new genus of marine bacteria is included among halophiles that are capable of producing extracellular hydrolytic enzymes, according to the research by Ardakani et al. [69], who isolated extracellular hydrolytic enzymes from the water and sediments of the Persian Gulf, in that site the isolation of bacteria that produced enzymes belong to the *Pseudoalteromonas* genera, and the activities include amylase, protease, and lipase.

During 2013, studies were made on enzyme-producing halophilic archaea capable of synthesizing two new alcohol-dehydrogenases, amylase and a thermostable halo-alkaliphile α-amylase; the producing organisms were identified as *Haloferax volcanii, Natrialba aegyptiaca* and *Halorubrum xinjiangense*, respectively. Hagaggi et al. [87] reported the isolation of the extremely halophilic archaea *Natrialba aegyptiaca*, from a salty soil near Aswan in Egypt. This organism is capable of producing an extracellular halophilic amylase that digests raw starch; therefore the enzyme may be used to efficiently process different vegetable sources. Moshfegh et al. [71] isolated a thermostable halo-alkaliphile α-amylase from an archaea located in the salty water of the Urmia Lake, which lies in northeast Iran. The producing organism was identified as *Halorubrum xinjiangense*, based on its morphological, biochemical, and molecular properties. Nigam et al. (2013) [72] tested the alkaline proteases produced by halophilic bacteria isolated from the Sambar Lake in Rajasthan for keratolytic activity. Moreno et al. [88] have shown that some microorganisms from hypersaline environments in Spain are able to produce hydrolytic enzymes; these have been related to the genera *Salinivibrio, Halomonas, Chromohalobacter, Bacillus-Salibacillus, Salinicoccus, Marinococcus, Halorubrum, Haloarcula, Halobacterium, Salicola, Salinibacter*, and *Pseudomonas*.

The bacteria isolated from a saline lake in Iran produced lipases, these bacteria belonging to the genera *Salicola, Halovibrio, Halomonas, Oceanobacillus, Thalassobacillus, Halobacillus, Virgibacillus, Gracilibacillus, Salinicoccus*, and *Piscibacillus*. On the other hand, sediments of deep waters in China have been found to contain amylase-producing organisms from the genera *Alcanivorax, Bacillus, Cobetia, Halomonas, Methylarcula, Micrococcus, Myroides, Paracoccus, Planococcus, Pseudomonas, Psychrobacter, Sporosarcina, Sufflavibacter*, and *Wangia*. In a desert in Chile, enzymes with DNase activity were related to the genera *Bacillus, Halobacillus, Pseudomonas, Halomonas*, and *Staphylococcus* [89].

A halo-alkaliphile, thermostable extracellular protease was reported by Selim et al. [74] produced by *Natronolimnobius innermongolicus* WN18 (HQ658997), an organism that belongs to the genus *Natronolimnobius* and was isolated from the sodium lake of An-Natrun, Egypt. Another halophile studied in the same year was the marine bacterium *Zunongwangia profunda,* due to its production of a new α-amylase resistant to low temperatures and tolerant to high concentrations of NaCl (4 M) [90].

Gupta et al. [75] reported a halo-alkaliphile isolated from a soil sample collected from the Sambhar Lake in Rajasthan, northern India, which produced an extracellular alkaline protease; the results of the analysis of gene 16S rRNA showed a 98% match with *Halobiforma* sp. Del Campo et al. [91] and Kumar and Khare [92] used fermentation in a solid medium to produce an esterase from halophilic archaea (*Natronococcus* sp. TC6, *Halobacterium* sp. NRC-1, and *Haloarcula marismortui*). These authors also optimized the production and nano-immobilization of *Marinobacter* sp. for an efficient hydrolysis of starch.

The *Kocuria* is mentioned as an example of a genus capable of producing extracellular amylases [77]. On the other hand, archaea *Halobacterium* sp. was isolated from samples of fermented fish and was considered a strong source of halophilic protease [93]; the bacterium *Bacillus licheniformis* isolated from sea water and sediments in Alexandria Eastern Harbor, Egypt, together with *Bacillus subtilis* isolated from the salt mines in Karak, Pakistan, were found to be producers of extracellular amylases and proteases, respectively [76, 94].

Dumorné et al. [95] stated that that the halophiles *Acinetobacter, Haloferax, Halobacterium, Halorhabdus, Marinococcus, Micrococcus, Natronococcus, Bacillus, Halobacillus,* and *Halothermothrix* produce extremozymes such as xylanases, amylases, proteases, and lipases. Halophile bacterium *Idiomarina* produces two extracellular proteases and was isolated in Badab-Sourt, Irán [79]. The same year, Hosseini et al. [96] described the isolation of bacteria capable of nitrite reduction that belonged to five different genera: *Bacillus, Halobacillus, Idiomarina, Oceanobacillus,* and *Virgibacillus,* in a paper regarding denitrifying halophile bacteria. Isolates capable of producing nitrate reductase were found among the genera *Halobacillus* and *Halomonas*. Another study on soils was carried out by Bhatt et al. [78], who isolated halo-alkaliphile bacteria from the saline desert soil in Little Rann of Kutch, India. Phylogenetic analysis indicated that isolates belong to phylum Firmicutes, which comprises lower G + C Gram-positive bacteria of different genera. Most of the halophilic isolates produced proteases (30% of isolates), followed by cellulases (24% isolates), CMCases (24% of isolates), and amylases (20% of isolates).

2.4. Physicochemical parameters and kinetic properties of extremozymes from halophilic microorganisms

Halophilic enzymes have specific mechanisms for solubility at high salt concentrations, such as a highly negative superficial charge given by carboxylic groups that depend on high salt concentrations to remain soluble. Halophilic archaea are known to secrete active proteases at high concentrations of NaCl (4 M), and to accumulate high concentrations of KCl in their cytoplasm in order to face osmotic stress, while maintaining the conformation of their proteins. The study made by Akolkar and Desai [97] suggests that proteases from haloarchaea may be active and stable in the presence of osmolytes different from NaCl/KCl at different degrees, as shown by the kinetics and thermodynamic analyses of casein hydrolysis produced by *Halobacterium* sp., in the presence of a compatible solute (sodium glutamate). In 2012, Zhang et al. [98] demonstrated that NaCl may improve the thermal stability of enzymes, and the presence of NaCl or KCl increases enzymatic activity 10-fold, approximately; this agrees with other investigations that demonstrate that enzyme activity depends on the concentration of NaCl or KCl, as well as on the substrate, pH, and presence of ions. Moshfegh et al. [71] demonstrated that the concentration of NaCl 4 M or KCl 4.5 M determines the maximum activity of halo-alkaliphile α-amylase produced by archaea *Halorubrum xinjiangense*, besides improving its thermal stability. In a similar manner, Selim et al. [74] showed that the protease activity of *Natronolimnobius innermongolicus* depends on high concentrations of salt to remain active and stable. Proteases purified by Faghihi et al. [79] increased their activity in the presence of metallic ions such as Mn^{2+} and Cu^{2+}, while decreasing activity when exposed to Hg^{2+} and Fe^{2+}. Both proteases were strongly inhibited by SDS, while DDT, EDTA, and 2-mercaptoethanol may stimulate their activity.

The affinity of an enzyme to hydrolyze a substrate is determined by the Michaelis constant (K_m), which is the concentration of substrate at which the reaction velocity is half the maximum velocity (V_{max}). V_{max} is the maximum velocity when the system is saturated with substrate. The value of *Km* is a measure of the enzyme-substrate affinity, and at the moment there are very few determinations made on extremozymes from halophilic organisms. **Table 2** shows the values of kinetic constants for extremozymes currently available.

Nowadays, recombinant DNA techniques and genetic engineering are used to obtain customized extremozymes to be used for specific purposes, greatly improving their catalytic ability, as demonstrated by Kui et al. [99] with the expression of genes from extremozyme β-1,4-xylanase, which was cloned from *Nesterenkonia xinjiangensis* and expressed in *Escherichia coli*. This enzyme was thermostable, retaining more than 80% of the initial activity after incubation at 60°C for 1 h, and more than 40% activity at 90°C for 15 min. In the same way, Qin et al. [90] cloned a novel gene that codifies a new α-amylase, which is active at low temperatures and tolerant to salt (AmyZ), from the marine bacterium *Zunongwangia profunda*, this protein was also expressed in *Escherichia coli*. It was observed that AmyZ is one of the few α-amylases that tolerate both low temperatures and high salinity, which makes it a potential candidate for research in basic and applied biology.

2.5. Halophile extremozyme applications

As mentioned above, halophiles are good sources of several extremozymes, and among them hydrolases have been the most studied, mainly amylases, proteases, lipases, xylanases, cellulases, and DNases. Some extremozymes from halophiles exhibit extraordinary biochemical properties, which show the potential for industrial applications. It has been demonstrated that extremozymes derived from halophiles are able to function under harsh conditions and remain

Microorganism	Extremonzyme	km (mg/mL)	V_{max}	Reference
Nesterenkonia xinjiangensis	Xilanasa	16.08	45.66 μmol/min-mg	[99]
Bacillus sp.	Celulasa	3.18		[98]
Halorubrum xinjiangense	α-amilasa	3.8	12.4 U/mg	[71]
Aspergillus gracilis	α-amilasa	6.33	8.36 U/mg	[87]
Kocuria sp.	Amilasa	3.0	90.09 U/ml	[77]

Table 2. Kinetics parameters of extreme-enzymes produced by halophiles microorganims.

stable and active with different properties than conventional enzymes, offering opportunities in several applications such as environmental bioremediation, food processing, and residual water treatment. Recent research points out the application of halophilic extremozymes in the production of biofuels. Since several halophiles are also alkaliphiles, their enzymes are of interest for the textile and detergent industries, and some have been explored as raw materials in the production of commercial enzymes, particularly proteases and amylases [5, 100–105].

3. Conclusions

Halophile microorganisms have the ability to adapt to or tolerate stress caused by salinity by accumulating osmolytes. Halophilic microorganisms have several biotechnological applications, in recent years, uses of halophilic microorganisms have significantly increased. Many enzymes, stabilizers, and valuable compounds from halophiles may present advantages for the development of biotechnological production processes. Halophiles are the most probable source of extremozymes, since them also capable of tolerating alkaline pH and high temperatures.

Acknowledgements

The authors thank the support granted by PROFAPI-ITSON Projects 2018-1076 and 2018-1169 for the realization of the present investigation. MS Vásquez-Murrieta appreciates the scholarships of Comisión de Operación y Fomento de Actividades Académicas (COFAA), Estímulos al Desempeño de los Investigadores (EDI-IPN), and Sistema Nacional de Investigadores (SNI-CONACyT). B Román-Ponce thanks for the Postdoctoral fellowships awarded by Consejo Nacional de Ciencia y Tecnología (CONACyT).

Author details

Luis Alberto Cira-Chávez[1], Joseph Guevara-Luna[2], Marisela Yadira Soto-Padilla[3], Brenda Román-Ponce[2,4], María Soledad Vásquez-Murrieta[2] and María Isabel Estrada-Alvarado[1]*

*Address all correspondence to: maria.estrada@itson.edu.mx

1 Departamento de Biotecnología y Ciencias Alimentarias, Instituto Tecnológico de Sonora, Sonora, Mexico

2 Departamento de Microbiología, Escuela Nacional de Ciencias Biológicas, Instituto Politécnico Nacional. Prol. de Carpio y Plan de Ayala, Ciudad de Mexico, Mexico

3 Instituto de Ingeniería y Tecnología, Universidad Autónoma de Ciudad Juárez, Mexico

4 Departamento de Microbiología y Genética, Edificio Departamental, Salamanca, Spain

References

[1] DasSarma S, DasSarma P. Halophiles. In: eLS. Chichester: John Wiley & Sons, Ltd. p. 2012. DOI: 10.1002/9780470015902.a0000394.pub3

[2] Ollivier B, Caumette P, Garcia JL, Mah RA. Anaerobic bacteria from hypersaline environments. Microbiological Reviews. 1994;**58**:27-38

[3] Madigan MT, Martinko JM, Parker J. Brock Biology of Microorganisms. Upper Saddle River, NJ: Prentice Hall; 1997. p. 11

[4] Ventosa A, Arahal DR. Physico-chemical characteristics of hypersaline environments and their biodiversity. Extremophiles. 2009;**2**:247-262

[5] DasSarma S, DasSarma P. Halophiles and their enzymes: Negativity put to good use. Current Opinion in Microbiology. 2015;**25**:120-126. DOI: 10.1016/j.mib.2015.05.009

[6] Kushner DJ, Kamekura M. Physiology of halophilic eubacteria. In: Rodriguez-Valera R, editor. Halophilic Bacteria. Boca Raton, Fla: CRC Press; 1988

[7] Ventosa A. Unusual micro-organisms from unusual habitats: Hypersaline environments. In: Symposia-Society for General Microbiology. Cambridge: Cambridge University Press; 1999

[8] Ventosa A, de la Haba RR, Sánchez-Porro C, Papke RT. Microbial diversity of hypersaline environments: A metagenomic approach. Current Opinion in Microbiology. 2015;**25**:80-87. DOI: 10.1016/j.mib.2015.05.002

[9] La Cono V, Smedile F, Bortoluzzi G, Arcadi E, Maimone G, Messina E, et al. Unveiling microbial life in new deep-sea hypersaline Lake Thetis. Part I: Prokaryotes and environmental settings. Environmental Microbiology. 2011;**13**(8):2250-2268

[10] Oren A. Ecology of halophiles. In: Horikoshi K, Antranikian G, Bull AT, Robb FT, Stetter KO, editors. Extremophiles Handbook. Japan: Springer. DOI: 10.1007/978-4-431-53898-1_3.2

[11] Yakimov MM, La Cono V, Spada GL, Bortoluzzi G, Messina E, Smedile F, et al. Microbial community of the deep-sea brine Lake Kryos seawater–brine interface is active below the chaotropicity limit of life as revealed by recovery of mRNA. Environmental Microbiology. 2015;**17**(2):364-382. DOI: 10.1111/1462-2920

[12] Antón J, Oren A, Benlloch S, Rodríguez-Valera F, Amann R, Rosselló-Mora R. *Salinibacter ruber* gen. nov., sp. nov., a novel, extremely halophilic member of the bacteria from saltern crystallizer ponds. International Journal of Systematic and Evolutionary Microbiology. 2002;**52**(2):485-491. DOI: 10.1099/00207713-52-2-485

[13] Yeon SH, Jeong WJ, Park JS. The diversity of culturable organotrophic bacteria from local solar salterns. Journal of Microbiology. 2005;**43**(1):1-10

[14] Lim JM, Jeon CO, Kim CJ. *Bacillus taeanensis* sp. nov., a halophilic gram-positive bacterium from a solar saltern in Korea. International Journal of Systematic and Evolutionary Microbiology. 2006;**56**(12):2903-2908. DOI: 10.1099/ijs.0.64036-0

[15] Pappa A, Sánchez-Porro C, Lazoura P, Kallimanis A, Perisynakis A, Ventosa A, et al. *Bacillus halochares* sp. nov., a halophilic bacterium isolated from a solar saltern. International Journal of Systematic and Evolutionary Microbiology. 2010;**60**(6):1432-1436. DOI: 10.1099/ijs.0.014233-0

[16] Maturrano L, Santos F, Rosselló-Mora R, Antón J. Microbial diversity in Maras salterns, a hypersaline environment in the Peruvian Andes. Applied and Environmental Microbiology. 2006;**72**(6):3887-3895. DOI: 10.1128/AEM.02214-05

[17] Baati H, Amdouni R, Gharsallah N, Sghir A, Ammar E. Isolation and characterization of moderately halophilic bacteria from Tunisian solar saltern. Current Microbiology. 2010;**60**(3):157-161. DOI: 10.1007/s00284-009-9516-6

[18] Shrivastava P, Kumar R. Soil salinity: A serious environmental issue and plant growth promoting bacteria as one of the tools for its alleviation. Saudi Journal of Biological Sciences. 2015;**22**(2):123-131. DOI: 10.1016/j.sjbs.2014.12.001

[19] Jamil A, Riaz S, Ashraf M, Foolad MR. Gene expression profiling of plants under salt stress. Critical Reviews in Plant Sciences. 2011;**30**(5):435-458. DOI: 10.1080/07352689.2011.605739

[20] Li WJ, Zhang YQ, Schumann P, Tian XP, Zhang YQ, Xu LH, et al. *Sinococcus qinghaiensis* gen. nov., sp. nov., a novel member of the order Bacillales from a saline soil in China. Inter national Journal of Systematic and Evolutionary Microbiology. 2006;**56**(6):1189-1192. DOI: 10.1099/ijs.0.64111-0

[21] Tian XP, Dastager SG, Lee JC, Tang SK, Zhang YQ, Park DJ, et al. *Alkalibacillus halophilus* sp. nov., a new halophilic species isolated from hypersaline soil in Xin-Jiang province, China. Systematic and Applied Microbiology. 2007;**30**(4):268-272. DOI: 10.1016/j.syapm.2006.08.003

[22] Chen YG, Cui XL, Li WJ, Xu LH, Wen ML, Peng Q, et al. *Salinicoccus salitudinis* sp. nov., a new moderately halophilic bacterium isolated from a saline soil sample. Extremophiles. 2008;**12**(2):197-203. DOI: 10.1007/s00792-007-0116-8

[23] Li WJ, Park DJ, Tang SK, Wang D, Lee JC, Xu LH, et al. *Nocardiopsis salina* sp. nov., a novel halophilic actinomycete isolated from saline soil in China. International Journal of Systematic and Evolutionary Microbiology. 2004;**54**(5):1805-1809. DOI: 10.1099/ijs.0.63127-0

[24] Meklat A, Bouras N, Riba A, Zitouni A, Mathieu F, Rohde M, et al. *Streptomonospora algeriensis* sp. nov., a halophilic actinomycete isolated from soil in Algeria. Antonie Van Leeuwenhoek. 2014;**106**(2):287-292. DOI: 10.1007/s10482-014-0195-3

[25] Martínez-Cánovas MJ, Quesada E, Martínez-Checa F, del Moral A, Bejar V. *Salipiger mucescens* gen. nov., sp. nov., a moderately halophilic, exopolysaccharide-producing bacterium isolated from hypersaline soil, belonging to the α-Proteobacteria. International Journal of Systematic and Evolutionary Microbiology. 2004;**54**(5):1735-1740. DOI: 10.1099/ijs.0.63166-0

[26] Martínez-Checa F, Quesada E, Martínez-Cánovas MJ, Llamas I, Bejar V. *Palleronia marisminoris* gen. nov., sp. nov., a moderately halophilic, exopolysaccharide-producing bacterium belonging to the 'Alphaproteobacteria', isolated from a saline soil. International

Journal of Systematic and Evolutionary Microbiology. 2005;**55**(6):2525-2530. DOI: 10.1099/ijs.0.63906-0

[27] Larsen H. Ecology of hypersaline environments. Developments in Sedimentology. 1980;**28**:23-39. DOI: 10.1016/S0070-4571(08)70227-9

[28] Jakobsen TF, Kjeldsen KU, Ingvorsen K. *Desulfohalobium utahense* sp. nov., a moderately halophilic, sulfate-reducing bacterium isolated from Great Salt Lake. International Journal of Systematic and Evolutionary Microbiology. 2006;**56**(9):2063-2069. DOI: 10.1099/ijs.0.64323-0

[29] Almeida-Dalmet S, Sikaroodi M, Gillevet PM, Litchfield CD, Baxter BK. Temporal study of the microbial diversity of the north arm of Great Salt Lake, Utah, US. Microorganisms. 2015;**3**(3):310-326. DOI: 10.3390/microorganisms3030310

[30] Demergasso C, Casamayor EO, Chong G, Galleguillos P, Escudero L, Pedrós-Alió C. Distri bution of prokaryotic genetic diversity in athalassohaline lakes of the Atacama Desert, northern Chile. FEMS Microbiology Ecology. 2004;**48**(1):57-69. DOI: 10.1016/j.femsec. 2003.12.013

[31] Rodríguez-Valera F. Characteristics and microbial ecology of hypersaline environments. In: Rodriguez-Valera F, editor. Halophilic Bacteria. Vol. 1. Boca Raton, FL: CRC Press. pp. 3-30

[32] McGenity TJ, Oren A. Life in saline environments. In: Bell EM, editor. Life at Extremes. Environments, Organisms and Strategies for Survival. UK: CABI International; 2012. pp. 402-437

[33] Al-Karablieh N. Antimicrobial activity of *Bacillus persicus* 24-DSM isolated from dead sea mud. The Open Microbiology Journal. 2017;**11**:372. DOI: 10.2174/1874285801711010372

[34] Ionescu D, Siebert C, Polerecky L, Munwes YY, Lott C, Häusler S, et al. Microbial and chemical characterization of underwater freshwater springs in the Dead Sea. PLoS One. 2012;**7**(6):e38319. DOI: 10.1371/journal.pone.0038319

[35] Bodaker I, Sharon I, Suzuki MT, Feingersch R, Shmoish M, Andreishcheva E, et al. Comparative community genomics in the Dead Sea: An increasingly extreme environment. The ISME Journal. 2010;**4**(3):399. DOI: 10.1038/ismej.2009.141

[36] van der Wielen PW, Bolhuis H, Borin S, Daffonchio D, Corselli C, Giuliano L, et al. The enigma of prokaryotic life in deep hypersaline anoxic basins. Science. 2005;**307**(5706):121-123. DOI: 10.1126/science.1103569

[37] Grant WD, Jones BE. Alkaline environments. Encyclopedia of Microbiology. 2000;**1 A-C**: 126-133

[38] Grant WD. Introductory chapter: Half a lifetime in soda lakes. In: Halophilic Micro organisms. Vol. 1. Berlin, Heidelberg: Springer; 2004. pp. 17-31. DOI: 10.1007/978-3-662-07 656-9

[39] Ma Y, Zhang W, Xue Y, Zhou P, Ventosa A, Grant WD. Bacterial diversity of the inner Mongolian Baer soda Lake as revealed by 16S rRNA gene sequence analyses. Extremophiles. 2004;**8**(1):45-51. DOI: 10.1007/s00792-003-0358-z

[40] Rees HC, Grant WD, Jones BE, Heaphy S. Diversity of Kenyan soda lake alkaliphiles assessed by molecular methods. Extremophiles. 2004;**8**(1):63-71. DOI: 10.1007/s00792-003-0361-4

[41] Mesbah NM, Hedrick DB, Peacock AD, Rohde M, Wiegel J. *Natranaerobius thermophilus* gen. nov., sp. nov., a halophilic, alkalithermophilic bacterium from soda lakes of the Wadi an Natrun, Egypt, and proposal of *Natranaerobiaceae* fam. nov. and *Natranaerobiales* ord. nov. International Journal of Systematic and Evolutionary Microbiology. 2007;**57**(11):2507-2512. DOI: 10.1099/ijs.0.65068-0

[42] Dimitriu PA, Pinkart HC, Peyton BM, Mormile MR. Spatial and temporal patterns in the microbial diversity of a meromictic soda lake in Washington state. Applied and Environmental Microbiology. 2008;**74**(15):4877-4888. DOI: 10.1128/AEM.00455-08

[43] Sorokin DY, Rusanov II, Pimenov NV, Tourova TP, Abbas B, Muyzer G. Sulfidogenesis under extremely haloalkaline conditions in soda lakes of Kulunda steppe (Altai, Russia). FEMS Microbiology Ecology. 2010;**73**(2):278-290. DOI: 10.1111/j.1574-6941.2010.00901.x

[44] Sorokin DY, Detkova EN, Muyzer G. Sulfur-dependent respiration under extremely haloalkaline conditions in soda lake 'acetogens' and the description of *Natroniella sulfidigena* sp. nov. FEMS Microbiology Letters. 2011;**319**(1):88-95. DOI: 10.1111/j.1574-6968.2011.02272.x

[45] Blum JS, Kulp TR, Han S, Lanoil B, Saltikov CW, Stolz JF, et al. *Desulfohalophilus alkaliarsenatis* gen. nov., sp. nov., an extremely halophilic sulfate-and arsenate-respiring bacterium from Searles Lake, California. Extremophiles. 2012;**16**(5):727-742. DOI: 10.1007/s00792-012-0468-6

[46] Sorokin DY, Tourova TP, Sukhacheva MV, Muyzer G. *Desulfuribacillus alkaliarsenatis* gen. nov. sp. nov., a deep-lineage, obligately anaerobic, dissimilatory sulfur and arsenate-reducing, haloalkaliphilic representative of the order Bacillales from soda lakes. Extremophiles. 2012;**16**(4):597-605. DOI: 10.1007/s00792-012-0459-7

[47] Sorokin DY, Chernyh NA, Poroshina MN. *Desulfonatronobacter acetoxydans* sp. nov.: A first acetate-oxidizing, extremely salt-tolerant alkaliphilic SRB from a hypersaline soda lake. Extremophiles. 2015;**19**(5):899-907. DOI: 10.1007/s00792-015-0765-y

[48] Sarwar MK, Azam I, Iqbal T. Biology and applications of halophilic bacteria and archaea: A review. Electronic Journal of Biology. 2015;**11**(3):98-103. ISSN: 1860-3122

[49] Oren A. Bioenergetic aspects of halophilism. Microbiology and Molecular Biology Reviews. 1999;**63**:334-340. DOI: 10.92-2172/99/$04.0010

[50] Yan N, Marschner P, Cao W, Zuo C, Qin W. Influence of salinity and water content on soil microorganisms. The International Soil and Water Conservation Research (ISWCR). 2015:3316-3323. DOI: 10.1016/j.iswcr.2015.11.003

[51] Edbeib MF, Wahab RA, Huyop F. Halophiles: Biology, adaptation, and their role in decontamination of hypersaline environments. World Journal of Microbiology and Biotechnology. 2016;**32**:135. DOI: 10.1007/s11274-016-2081-9

[52] Roberts MF. Organic compatible solutes of halotolerant and halophilic microorganisms. Saline Systems. 2005;**1**:5. DOI: 10.1186/1746-1448-1-5

[53] Oren A. Microbial life at high salt concentrations: Phylogenetic and metabolic diversity. Saline Systems. 2008;**4**:2. DOI: 10.1186/1746-1448-4-2

[54] Yin J, Chen JC, Wu Q, Chen GQ. Halophiles, coming stars for industrial biotechnology. Biotechnology Advances | Industrial Biotechnology: Tools and and Applications. 2015;**33**:1433-1442. DOI: 10.1016/j.biotechadv.2014.10.008

[55] Hanelt I, Muller V. Molecular mechanisms of adaptation of the moderately halophilic bacterium *Halobacillus halophilus* to its environment. Lifestyles. 2013;**3**:234-243. DOI: 10.3390/life3010234

[56] Sharma A, Vaishnav A, Jamali H, Kumar SA, Saxena AK, Kumar SA. Halophilic bacteria: Potential bioinoculants for sustainable agriculture and environment management under salt stress. In: Plant-Microbe Interaction: An Approach to Sustainable Agriculture. Singapore: Springer; 2016. pp. 297-325. DOI: 10.1007/978-981-10-2854-0_14

[57] Kunte HJ, Trüper HG, Stan-Lotter H. Halophilic microorganisms. In: Astrobiology. Berlin, Heidelberg: Springer; 2002. pp. 185-200. DOI: 10.1007/978-3-642-59,381-9_13

[58] Lanyi JK. Salt-dependent properties of proteins from extremely halophilic bacteria. Bacteriological Reviews. 1974;**38**:272-290

[59] Oren A. Diversity of halophiles. In: Horikoshi K, Antranikian G, Bull AT, Robb FT, Stetter KO, editors. Extremophiles Handbook. Japan: Springer; 2011. pp. 309-325. DOI: 10.1007/978-4-431-53,898-1_3.2

[60] Imhoff JF, Rodriguez-Valera F. Betaine is the main compatible solute of halophilic eubacteria. Journal of Bacteriology. 1984;**160**(1):478-479. DOI: 0021-9193/84/100478-02$02.00/0

[61] Robert MF. Osmodaption and osmoregulation in archaea: Update 2004. Frontiers in Bioscience. 2004;**9**:1999-2019. DOI: 10.2741/1366

[62] Ma Y, Galinski EA, Grant WD, Oren A, Ventosa A. Halophiles 2010: Life in saline environments. Applied and Environmental Microbiology. 2010;**76**:6971-6981. DOI: 10.1128/AEM.01868-10

[63] Delgado-García M, Valdivia-Urdiales B, Aguilar-González C, Contreras-Esquivel J, Rodríguez-Herrera R. Halophilic hydrolases as a new tool for the biotechnological industries. Journal of the Science of Food and Agriculture. 2012;**92**:2575-2580. DOI: 10.1002/jsfa.5860

[64] Mei Y, Liu H, Zhang S, Yang M, Hu C, Zhang J, et al. Effects of salinity on the cellular physiological responses of *Natrinema* sp. J7-2. PLoS One. 2017;**12**:e0184974. DOI: 10.1371/journal.pone.0184974

[65] Csonka LN. Physiological and genetic responses of bacteria to osmotic-stress. Microbiological Reviews. 1989;**53**:121-147. DOI: 0146-0749/89/010121-27$02.00/0

[66] Oren A. The bioenergetic basis for the decrease in metabolic diversity at increasing salt concentrations: Implications for the functioning of salt lake ecosystems. Hydrobiologia. 2001;**466**:61-72. DOI: 10.1023/A:1014557116838

[67] Joghee NN, Jayaraman G. Metabolomic characterization of halophilic bacterial isolates reveals strains synthesizing rare diaminoacids under salt stress. Biochimie. 2014;**102**:102-111. DOI: 10.1016/j.biochi.2014.02.015

[68] Vidyasagar M, Prakash S, Jayalakshmi SK, Sreeramulu K. Optimization of culture conditions for the production of halothermophilic protease from halophilic bacterium *Chromohalobacter* sp. World Journal of Microbiology and Biotechnology. 2007;**23**:655-662. DOI: 10.1111/j.1472-765X.2006.01980.x

[69] Ardakani MR, Poshtkouhian A, Amoozegar MA, Zolgharnein H. Isolation of moderately halophilic *Pseudoalteromonas* producing extracellular hydrolytic enzymes from Persian gulf. Indian Journal of Microbiology. 2012;**52**:94-98. DOI: 10.1007/s12088-011-0243-x

[70] Kumar S, Karan R, Kapoor S, Singh SP, Khare SK. Screening and isolation of halophilic bacteria producing industrially important enzymes. Brazilian Journal of Microbiology. 2012;**43**:1595-1603. DOI: 10.1590/S1517-83822012000400044

[71] Moshfegh M, Shahverdi AR, Zarrini G, Faramarzi MA. Biochemical characterization of an extracellular polyextremophilic α-amylase from the halophilic archaeon *Halorubrum xinjiangense*. Extremophiles. 2013;**17**:677-687. DOI: 10.1007/s00792-013-0551-7

[72] Nigam VK, Singhal P, Vidyarthi AS, Mohan MK, Ghosh P. Studies on keratinolytic activity of alkaline proteases from halophilic bacteria. International Journal of Pharma and Bio Sciences. **2013**(4):389-399

[73] Timpson LM, Liliensiek AK, Alsafadi D, Cassidy J, Sharkey MA, Liddell S, et al. A comparison of two novel alcohol dehydrogenase enzymes (ADH1 and ADH2) from the extreme halophile *Haloferax volcanii*. Applied Microbiology and Biotechnology. 2013;**97**:195-203. DOI: 10.1007/s00253-012-4074-4

[74] Selim S, Hagagy N, Aziz MA, El-Meleigy ES, Pessione E. Thermostable alkaline halophilic-protease production by *Natronolimnobius innermongolicus* WN18. Natural Product Research. 2014;**28**:1476-1479. DOI: 10.1080/14786419.2014.907288

[75] Gupta M, Aggarwal S, Navani NK, Choudhury B. Isolation and characterization of a protease-producing novel haloalkaliphilic bacterium *Halobiforma* sp. strain BNMIITR from Sambhar lake in Rajasthan, India. Annals of Microbiology. 2015;**65**:677-686. DOI: 10.1007/s13213-014-0906-z

[76] Ali N, Ullah N, Qasim M, Rahman H, Khan SN, Sadiq A, et al. Molecular characterization and growth optimization of halo-tolerant protease producing *Bacillus subtilis* strain BLK-1.5 isolated from salt mines of Karak, Pakistan. Extremophiles. 2016;**20**:395-402. DOI: 10.1007/s00792-016-0830-1

[77] Soto-Padilla MY, Gortáres-Moroyoqui P, Cira-Chávez LA, Levasseur A, Dendooven L, Estrada-Alvarado MI. Characterization of extracellular amylase produced by haloalkalophilic strain *Kocuria* sp. HJ014. International Journal of Environmental Health Research. 2016;**26**:396-404. DOI: 10.1080/09603123.2015.1135310

[78] Bhatt HB, Gohel SD, Singh SP. Phylogeny, novel bacterial lineage and enzymatic potential of haloalkaliphilic bacteria from the saline coastal desert of Little Rann of Kutch, Gujarat, India. 3 Biotech. 2018;**8**:53. DOI: 10.1007/s13205-017-1075-0

[79] Faghihi LS, Seyedalipour B, Riazi G, Ahmady-Asbchin S. Introduction of two halo-alkali-thermo-stable biocatalysts: Purification and characterization. Catalysis Letters. 2018;**148**:831-842. DOI: 10.1007/s10562-018-2295-6

[80] Sánchez-Porro C, Martin S, Mellado E, Ventosa A. Diversity of moderately halophilic bacteria producing extracellular hydrolytic enzymes. Journal of Applied Microbiology. 2003;**94**:295-300. DOI: 10.1046/j.1365-2672.2003.01834.x

[81] Karbalaei-Heidari HR, Ziaee AA, Schaller J, Amoozegar MA. Purification and characterization of an extracellular haloalkaline protease produced by the moderately halophilic bacterium, *Salinivibrio* sp. strain AF-2004. Enzyme and Microbial Technology. 2007;**40**:266-272. DOI: 10.1016/j.enzmictec.2006.04.006

[82] Rohban R, Amoozegar MA, Ventosa A. Screening and isolation of halophilic bacteria producing extracellular hydrolyses from Howz Soltan Lake, Iran. Journal of Industrial Microbiology & Biotechnology. 2009;**36**:333-340. DOI: 10.1007/s10295-008-0500-0

[83] Pérez D, Martín S, Fernández-Lorente G, Filice M, Guisán JM, Ventosa A, et al. A novel halophilic lipase, LipBL, showing high efficiency in the production of eicosapentaenoic acid (EPA). PLoS One. 2011;**6**:e23325. DOI: 10.1371/journal.pone.0023325

[84] Li X, Yu HY. Extracellular production of beta-amylase by a halophilic isolate, *Halobacillus* sp. LY9. Journal of Industrial Microbiology & Biotechnology. 2011;**38**:1837-1843. DOI: 10.1007/s10295-011-0972-1

[85] Shahbazi M, Karbalaei-Heidari HR. A novel low molecular weight extracellular protease from a moderately halophilic bacterium *Salinivibrio* sp. strain MS-7: Production and biochemical properties. Molecular Biology Research Communications. **2012**(1):45-56. DOI: 10.22099/mbrc.2012.576

[86] Jayachandra SY, Kumar A, Merley DP, Sulochana MB. Isolation and characterization of extreme halophilic bacterium *Salinicoccus* sp. JAS4 producing extracellular hydrolytic enzymes. Recent Research in Science and Technology. 2012;**4**:46-49

[87] Hagaggi NS, Hezayen FF, Abdul-Raouf UM. Production of an extracellular halophilic amylase from the extremely halophilic archaeon *Natrialba aegyptiaca* strain 40 T. Al-Azhar Bulletin of Science. 2013;**24**:93-107

[88] Moreno ML, Pérez D, García MT, Mellado E. Halophilic bacteria as a source of novel hydrolytic enzymes. Lifestyles. **2013**(3):38-51. DOI: 10.3390/life3010038

[89] Moreno ML, Piubeli F, Bonfá MR, García MT, Durrant LR, Mellado E. Analysis and characterization of cultivable extremophilic hydrolytic bacterial community in heavy-metal-contaminated soils from the Atacama Desert and their biotechnological potentials. Journal of Applied Microbiology. 2012;**113**:550-559

[90] Qin Y, Huang Z, Liu Z. A novel cold-active and salt-tolerant α-amylase from marine bacterium *Zunongwangia profunda*: Molecular cloning, heterologous expression and biochemical characterization. Extremophiles. 2014;**18**:271-281. DOI: 10.1007/s00792-013-0614-9

[91] Del Campo MM, Camacho RM, Mateos-Díaz JC, Müller-Santos M, Córdova J, Rodríguez JA. Solid-state fermentation as a potential technique for esterase/lipase production by halophilic archaea. Extremophiles. 2015;**19**:1121-1132. DOI: 10.1007/s00792-015-0784-8

[92] Kumar S, Khare SK. Chloride activated halophilic α-amylase from *Marinobacter* sp. EMB8: Production optimization and nanoimmobilization for efficient starch hydrolysis. Enzyme Research. 2015:1-9. DOI: 10.1155/2015/859485

[93] Chuprom J, Bovornreungroj P, Ahmad M, Kantachote D, Dueramae S. Approach toward enhancement of halophilic protease production by *Halobacterium* sp. strain LBU50301 using statistical design response surface methodology. Biotechnology Reports. 2016;**10**: 17-28. DOI: 10.1016/j.btre.2016.02.004

[94] Abel-Nabey HM, Farag AM. Production, optimization and characterization of extracellular amylase from halophilic *Bacillus lichineformis* AH214. African Journal of Biotechnology. 2016;**2016, 15**:670-683. DOI: 10.5897/AJB2015.15073

[95] Dumorné K, Camacho Córdova D, Astorga-Eló M, Renganathan P. Extremozymes: A potential source for industrial applications. Journal of Microbiology and Biotechnology. 2017;**27**(4):649-659. DOI: 10.4014/jmb.1611.11006

[96] Hosseini M, Al-Rubaye MTS, Fakhari J, Babaha F. Isolation and characterization of denitrifying halophilic bacteria from Bahr Al-Milh Salt Lake, Karbala, Iraq. Journal of Applied Biology & Biotechnology. 2018;**6**:32-36. DOI: 10.7324/JABB.2018.60406

[97] Akolkar AV, Desai AJ. Catalytic and thermodynamic characterization of protease from Halobacterium SP. SP1 (1). Research in Microbiology. 2010;**161**:355-362. DOI: 10.1016/j.resmic.2010.04.005

[98] Zhang G, Li S, Xue Y, Mao L, Ma Y. Effects of salts on activity of halophilic cellulase with glucomannanase activity isolated from alkaliphilic and halophilic *Bacillus* sp. BG-CS10. Extremophiles. 2012;**16**:35-43. DOI: 10.1007/s00792-011-0403-2

[99] Kui H, Luo H, Shi P, Bai Y, Yuan T, Wang Y, et al. Gene cloning, expression, and characterization of a thermostable xylanase from *Nesterenkonia xinjiangensis* CCTCC AA001025. Applied Biochemistry and Biotechnology. 2010;**162**:953-965. DOI: 10.1007/s12010-009-8815-5

[100] DasSarma P, Coker JA, Huse V, DasSarma S. Halophiles, industrial applications. In: Flickinger MC, editor. Encyclopedia of Industrial Biotechnology: Bioprocess, Biosepa ration, and Cell Technology. 2010. pp. 1-43

[101] Ma Y, Galinski EA, Grant WD, Oren A, Ventosa A. Halophiles 2010: Life in saline environments. Applied and Environmental Microbiology. 2010;**76**:6971-6981. DOI: 10.1128/AEM.01868-10

[102] Oren A. Industrial and environmental applications of halophilic microorganisms. Environmental Technology. 2010;**31**:825-834. DOI: 10.1080/09593330903370026

[103] Oliart-Ros RM, Manresa-Presas Á, Sánchez-Otero MG. Utilización de microorganismos de ambientes extremos y sus productos en el desarrollo biotecnológico. Ciencia UAT. 2016;**11**:79-90

[104] Waditee-Sirisattha R, Kageyama H, Takabe T. Halophilic microorganism resources and their applications in industrial and environmental biotechnology. AIMS Microbiology. 2016;**2**:42-54. DOI: 10.3934/microbiol.2016.1.42

[105] Javed S, Azeem F, Hussain S, Rasul I, Siddique MH, Riaz M, et al. Bacterial lipases: A review on purification and characterization. Progress in Biophysics and Molecular Biology. 2017;**132**:23-34. DOI: 10.1016/j.pbiomolbio.2017.07.014

[106] Lucena T, Arahal DR, Ruvira MA, Navarro-Torre S, Mesa J, Pajuelo E, et al. *Vibrio palustris* sp. nov. and *Vibrio spartinae* sp. nov., two novel members of the Gazogenes clade, isolated from salt-marsh plants (*Arthrocnemum macrostachyum* and *Spartina maritima*). International Journal of Systematic and Evolutionary Microbiology. 2017;**67**(9):3506-3512. DOI: 10.1099/ijsem.0.002155

[107] Bheemaraddi MC, Patil S, Shivannavar CT, Gaddad SM. Isolation and characterization of *Paracoccus* sp. GSM2 capable of degrading textile azo dye reactive violet 5. The Scientific World Journal. 2014:410704. DOI: 10.1155/2014/410704

[108] Zhang X, Gao J, Zhao F, Zhao Y, Li. Characterization of a salt-tolerant bacterium *Bacillus* sp. from a membrane bioreactor for saline wastewater treatment. Journal of Environmental Sciences. 2014;**26**(6):1369, 1374. DOI: 10.1016/S1001-0742(13)60613-0

[109] Fidalgo C, Martins R, Proença DN, Morais PV, Alves A. Henriques: *Zunongwangia endophytica* sp. nov., an endophyte isolated from the salt marsh plant, *Halimione portulacoides,* and emended description of the genus *Zunongwangia*. International Journal of Systematic and Evolutionary Microbiology. 2017;**67**(8):3004, 3009. DOI: 10.1099/ijsem.0.002069

[110] Maldonado LA, Fenical W, Jensen PR, Kauffman CA, Mincer TJ, Ward AC, et al. Salinispora arenicola gen. nov., sp. nov. and *Salinispora tropica* sp. nov., obligate marine actinomycetes belonging to the family Micromonosporaceae. International Journal of Systematic and Evolutionary Microbiology. 2005;**55**(5):1759-1766

[111] Ahmed L, Jensen PR, Freel KC, Brown R, Jones AL, Kim BY, et al. *Salinispora pacifica* sp. nov., an actinomycete from marine sediments. Antonie Van Leeuwenhoek. 2013;**103**(5):1069-1078. DOI: 10.1007/s10482-013-9886-4

[112] Bibi F, Chung EJ, Khan A, Jeon CO, Chung YR. *Martelella endophytica* sp. nov., an antifungal bacterium associated with a halophyte. International Journal of Systematic and Evolutionary Microbiology. **63**(8):2914-2919. DOI: 10.1099/ijs.0.048785-0

[113] Qin S, Bian GK, Tamura T, Zhang YJ, Zhang WD, Cao CL, et al. *Streptomyces halophytocola* sp. nov., an endophytic actinomycete isolated from the surface-sterilized stems of a coastal halophyte *Tamarix chinensis* lour. International Journal of Systematic and Evolutionary Microbiology. 2013;**63**(8):2770-2775. DOI: 10.1099/ijs.0.047456-0

[114] Bibi F, Jeong JH, Chung EJ, Jeon CO, Chung YR. *Labrenzia suaedae* sp. nov., a marine bacterium isolated from a halophyte, and emended description of the genus Labrenzia.

International Journal of Systematic and Evolutionary Microbiology. **64**(4):1116-1122. DOI: 10.1099/ijs.0.047456-0

[115] Roman-Ponce B, Wang D, Vásquez-Murrieta MS, Chen WF, Estrada-de los Santos P, Sui XH, et al. *Kocuria arsenatis* sp. nov., an arsenic-resistant endophytic actinobacterium associated with *Prosopis laegivata* grown on high-arsenic-polluted mine tailing. International Journal of Systematic and Evolutionary Microbiology. 2016;**66**(2):1027-1033. DOI: 10. 1099/ijsem.0.000830

[116] Abbas N, Hussain S, Azeem F, Shahzad T, Bhatti SH, Imran M, et al. Characterization of a salt resistant bacterial strain *Proteus* sp. NA6 capable of decolorizing reactive dyes in presence of multi-metal stress. World Journal of Microbiology and Biotechnology. 2017;**32**(11):181. DOI: 10.1007/s11274-016-2141-1

[117] Sorokin DY, Chernyh NA. 'Candidatus Desulfonatronobulbus propionicus': A first haloalkaliphilic member of the order Syntrophobacterales from soda lakes. Extremophiles. 2016;**20**(6):895-901. DOI: 10.1007/s00792-016-0881-3

[118] Krishnan R, Menon RR, Busse HJ, Tanaka N, Krishnamurthi S, Rameshkumar N. *Novosphingobium pokkalii* sp. nov, a novel rhizosphere-associated bacterium with plant beneficial properties isolated from saline-tolerant pokkali rice. Research in Microbiology. 2017;**168**(2):113-121. DOI: 10.1016/j.resmic.2016.09.001

[119] León MJ, Sánchez-Porro C, Ventosa A. *Marinobacter aquaticus* sp. nov., a moderately halophilic bacterium from a solar saltern. International Journal of Systematic and Evolutionary Microbiology. 2017;**67**(8):2622-2627. DOI: 10.1099/ijsem.0.001984

[120] Yan J, Li Y, Yan H, Chen WF, Zhang X, Wang ET, et al. *Agrobacterium salinitolerans* sp. nov., a saline–alkaline-tolerant bacterium isolated from root nodule of Sesbania cannabina. International Journal of Systematic and Evolutionary Microbiology. 2017;**67**(6):1906-1911. DOI: 10.1099/ijsem.0.001885

[121] Zhao GY, Zhao LY, Xia ZJ, Zhu JL, Liu D, Liu CY, et al. *Salinicola tamaricis* sp. nov., a heavy-metal-tolerant, endophytic bacterium isolated from the halophyte tamarix chinensis lour. International Journal of Systematic and Evolutionary Microbiology. 2017;**67**(6):1813-1819. DOI: 10.1099/ijsem.0.001868

[122] Hou J, Zhao YJ, Zhu L, Cui HL. *Salinirubellus salinus* gen nov, sp. nov, isolated from a marine solar saltern. International Journal of Systematic and Evolutionary Microbiology. 2018;**68**(6):1874-1878. DOI: 10.1099/ijsem.0.002757

[123] Xia J, Ling SK, Wang XQ, Chen GJ, Du ZJ. *Aliifodinibius halophilus* sp. nov, a moderately halophilic member of the genus *Aliifodinibius*, and proposal of Balneolaceae fam nov. International Journal of Systematic and Evolutionary Microbiology. 2016;**66**(6):2225-2233. DOI: 10.1099/ijsem.0.001012

[124] Kjeldsen KU, Jakobsen TF, Glastrup J, Ingvorsen K. Desulfosalsimonas propionicica gen nov, sp. nov, a halophilic, sulfate-reducing member of the family Desulfobacteraceae isolated from a salt-lake sediment. International Journal of Systematic and Evolutionary Microbiology. 2010;**60**(5):1060-1065. DOI: 10.1099/ijs.0.014746-0

[125] Makhdoumi-Kakhki A, Amoozegar MA, Ventosa A. *Salinibacter iranicus* sp. nov and *Salinibacter luteus* sp. nov, isolated from a salt lake, and emended descriptions of the genus *Salinibacter* and of *Salinibacter ruber*. International Journal of Systematic and Evolutionary Microbiology. 2012;**62**(7):1521-1527. DOI: 10.1099/ijs.0.031971-0

[126] Abdeljabbar H, Cayol JL, Hania WB, Boudabous A, Sadfi N, Fardeau ML. *Halanaerobium sehlinense* sp. nov, an extremely halophilic, fermentative, strictly anaerobic bacterium from sediments of the hypersaline lake Sehline Sebkha. International Journal of Systematic and Evolutionary Microbiology. 2013;**63**(6):2069-2074. DOI: 10.1099/ijs.0.040139-0

[127] Abdallah MB, Karray F, Mhiri N, Cayol JL, Tholozan JL, Alazard D, et al. Characterization of *Sporohalobacter salinus* sp. nov, an anaerobic, halophilic, fermentative bacterium isolated from a hypersaline lake. International Journal of Systematic and Evolutionary Microbiology. 2015;**65**(2):543-548. DOI: 10.1099/ijs.0.066845-0

[128] Oh YJ, Lee HW, Lim SK, Kwon MS, Lee J, Jang JY, et al. *Lentibacillus kimchii* sp. nov, an extremely halophilic bacterium isolated from kimchi, a Korean fermented vegetable. Antonie Van Leeuwenhoek. **109**(6):869-876. DOI: 10.1007/s10482-016-0686-5

[129] Han JR, Ling SK, Yu WN, Chen GJ, Du ZJ. *Marinobacter salexigens* sp. nov, isolated from marine sediment. International Journal of Systematic and Evolutionary Microbiology. 2017;**67**(11):4595-4600. DOI: 10.1099/ijsem.0.002337

[130] Lu DC, Xia J, Dunlap CA, Rooney AP, Du ZJ. *Gracilimonas halophila* sp. nov, isolated from a marine solar saltern. International Journal of Systematic and Evolutionary Microbiology. 2017;**67**(9):3251-3255. DOI: 10.1099/ijsem.0.002093

[131] Nikou MM, Ramezani M, Harirchi S, Makzoom S, Amoozegar MA, Fazeli SAS, et al. *Salinifilum* gen nov, with description of *Salinifilum proteinilyticum* sp. nov, an extremely halophilic actinomycete isolated from Meighan wetland, Iran, and reclassification of *Saccharopolyspora aidingensis* as *Salinifilum aidingensis* comb nov and *Saccharopolyspora ghardaiensis* as *Salinifilum ghardaiensis* comb nov. International Journal of Systematic and Evolutionary Microbiology. 2017;**67**(10):4221-4227. DOI: 10.1099/ijsem.0.002286

[132] Sorokin DY, Kublanov IV, Khijniak TV. *Natronospira proteinivora* gen nov, sp. nov, an extremely salt-tolerant, alkaliphilic gammaproteobacterium from hypersaline soda lakes. International Journal of Systematic and Evolutionary Microbiology. 2017;**67**(8):2604-2608. DOI: 10.1099/ijsem.0.001983

[133] Amoozegar MA, Makhdoumi-Kakhki A, Ramezani M, Nikou MM, SAS F, Schumann P, et al. *Limimonas halophila* gen nov, sp. nov, an extremely halophilic bacterium in the family Rhodospirillaceae. International Journal of Systematic and Evolutionary Microbiology. 2013;**63**(4):1562-1567. DOI: 10.1099/ijs.0.041236-0

[134] Guan TW, Zhao K, Xiao J, Liu Y, Xia ZF, Zhang XP, et al. *Brevibacterium salitolerans* sp. nov, an A*ctinobacterium* isolated from salt-lake sediment. International Journal of Systematic and Evolutionary Microbiology. 2010;**60**(12):2991-2995. DOI: 10.1099/ijs.0.020214-0

[135] Bian GK, Feng ZZ, Qin S, Xing K, Wang Z, Cao CL, et al. *Kineococcus endophytica* sp. nov, a novel endophytic actinomycete isolated from a coastal halophyte in Jiangsu, China. Antonie Van Leeuwenhoek. 2012;**102**(4):621-628. DOI: 10.1007/s10482-012-9757-4

[136] Shi W, Takano T, Liu S. *Anditalea andensis* gen nov, sp. nov, an alkaliphilic, halotolerant bacterium isolated from extreme alkali–saline soil. Antonie Van Leeuwenhoek. 2012;**102**(4):703-710. DOI: 10.1007/s10482-012-9770-7

[137] Choi EJ, Lee SH, Jung JY, Jeon CO. *Brevibacterium jeotgali* sp. nov, isolated from jeotgal, a traditional Korean fermented seafood. International Journal of Systematic and Evolutionary Microbiology. 2013;**63**(9):3430-3436. DOI: 10.1099/ijs.0.049197-0

[138] Xin L, Hui-Ying Y. Purification and characterization of an extracellular esterase with organic solvent tolerance from a halotolerant isolate, *Salimicrobium* sp. LY19. BMC Biotechnology;**13**(1):08. DOI: 10.1186/1472-6750-13-108

[139] Román-Ponce B, Li YH, Vásquez-Murrieta MS, Sui XH, Chen WF, Estrada-De Los Santos P, et al. *Brevibacterium metallicus* sp. nov, an endophytic bacterium isolated from roots of *Prosopis laegivata* grown at the edge of a mine tailing in Mexico. Archives of Microbiology. 2015;**197**(10):1151-1158. DOI: 10.1007/s00203-015-1156-6

[140] Oguntoyinbo FA, Cnockaert M, Cho GS, Kabisch J, Neve H, Bockelmann W, et al. *Halomonas nigrificans* sp. nov, isolated from cheese. International Journal of Systematic and Evolutionary Microbiology. 2017;**68**(1):371-376. DOI: 10.1099/ijsem.0.002515

Chapter 2

Thermodynamic Activity-Based Michaelis Constants

Anton Wangler, Mark Jonathan Bunse,
Gabriele Sadowski and Christoph Held

Additional information is available at the end of the chapter

http://dx.doi.org/10.5772/intechopen.80235

Abstract

The classical approach towards analysing the influence of co-solvents (i.e., cellular molecules that are chemically inert and do not act as reacting agents) on the Michaelis constants of enzyme-catalysed reactions is empirical. More precisely, reaction kinetics is usually mathematically modelled by fitting empirical parameters to experimental concentration vs. time data. In this chapter, a thermodynamic approach is presented that replaces substrate concentrations by thermodynamic activities of the substrates. This approach allows determining activity-based Michaelis constants. The advantage of such activity-based constants K_M^a over their concentration-based pendants K_M^{obs} is twofold: First, K_M^a is independent of any co-solvent added (while K_M^{obs} is not) as long as it does not directly interfere with the reaction mechanism (e.g., inhibitor or activator). Second, known K_M^a values allow predictions of Michalis constants for different enzymes and reactions under co-solvent influence. This is demonstrated for a pseudo-one-substrate peptide hydrolysis reaction as well as for more complex two-substrate alcohol dehydrogenase reactions.

Keywords: enzyme kinetics, thermodynamics, activity coefficient, co-solvent, ePC-SAFT

1. Introduction

Understanding the kinetics of enzyme-catalysed reactions is a key aspect not just in the field of biology but also of high relevance for biocatalysis in the industry as enzymes are highly suitable for the production of fine chemicals [1]. The advantage of enzyme catalysis is that high enantioselectivity [2, 3] can often be realised under mild reaction conditions (ambient temperature and pressure).

Key properties for the study of enzyme-catalysed reactions are reaction yield and reaction kinetics. In case of the reaction yield, thermodynamic states an independence of the equilibrium position from the catalyst involved (as long as the catalyst concentration is low) [4–7]. In contrast, reaction kinetics strongly depends on the catalyst [8, 9]. This means that different enzymes used for the same reaction will cause different kinetic profiles for the considered reaction; this is represented by the experimental (concentration-based) Michaelis constant K_M^{obs} and the catalytic constant k_{cat}. These constants are thus enzyme-specific. Even more, the presence of co-solvents (i.e., chemically inert substances that do not act as metabolites) such as organic and inorganic compounds, salts and polymers might significantly influence such kinetic constants. In literature, the influence of diverse co-solvents on kinetics of a large amount of different enzyme-catalysed reactions is reported [7, 10–13]. It is common practice to empirically describe the co-solvent effects on the kinetic parameters; this requires a co-solvent-dependent consideration of enzyme kinetics. Further, it is discussed whether co-solvent-induced changes on K_M^{obs} are directly related to interactions between co-solvent and the catalytic centre or other parts of the enzyme. These former accepted relations have been recently revised in the publications of Grosch et al. [10], Pleiss [14, 15] and Wangler et al. [7]. These recent works suggested an approach, which is independent of the enzyme itself. In their approach, K_M^{obs} is influenced by co-solvent-substrate interactions caused by co-solvent-induced non-covalent molecular interactions between substrate and reaction medium, which the co-solvent is part of. By changing the perspective from co-solvent-enzyme interactions to co-solvent-substrate interactions, a new activity-based Michaelis constant K_M^a was proposed, which is based on thermodynamic activities of the substrates under co-solvent influence. The advantage of this treatment is that K_M^a is independent of any kind or concentration of co-solvents present in the reaction mixture. This is even more impressive as these recent works neglect co-solvent-enzyme interactions in order to obtain co-solvent independent values for K_M^a. Further, the advantage of such activity-based treatment over the concentration-based approach is to establish a non-empirical method towards predicting and understanding co-solvent effects on the Michaelis constants without the need of experimental kinetic data of reaction mixtures containing co-solvents. This method requires activity coefficients γ of the substrates. These activity coefficients describe the molecular interactions in the reaction mixture, and they can be predicted with different thermodynamic models, e.g., NRTL [16], UNIFAC [17] or ePC-SAFT [18].

In this chapter, an approach is presented to determine K_M^a values based on K_M^{obs} values of the neat (co-solvent-free) reaction system and the activity coefficients of the substrates. The considered reactions are the hydrolysis of N-succinyl-L-phenylalanine-p-nitroanilide (SPNA) catalysed by the enzyme α-chymotrypsin (α-CT) and a two-substrate reaction, namely the reduction of acetophenone (ACP) catalysed by alcohol dehydrogenase 270 (ADH 270) and by alcohol dehydrogenase 200 (ADH 200). Determined K_M^a values under neat conditions were used to predict the co-solvent influence on K_M^{obs} values of the reactions under consideration. These predicted values were finally compared to experimental data to validate this approach.

2. Pseudo-one-substrate reactions

2.1. Theoretical background

2.1.1. Concentration-based approach

Examples for one-substrate reactions are isomerase reactions where one substrate is converted to another without any change to the chemical composition of the molecule. Examples can be found in glycolysis, one being the reversible conversion of 3-phosphoglycerate (substrate S) to 2-phosphoglycerate (product P) catalysed by phosphoglycerate mutase (enzyme E). The general reaction scheme of a one-substrate reaction is given in Eq. (1).

$$E + S \rightleftharpoons ES \rightarrow E + P \tag{1}$$

The kinetics of the reaction according to Eq. (2) is commonly described by the Michaelis-Menten equation including the reaction rate v, the maximum reaction rate v_{max}, the Michaelis constant K_M^{obs} and the substrate molality m_S in mol/kg$_{water}$.

$$v = \frac{v_{max} \cdot m_S}{K_M^{obs} + m_S} \tag{2}$$

Eq. (2) is visualised exemplary by plotting of v over m_S in **Figure 1**.

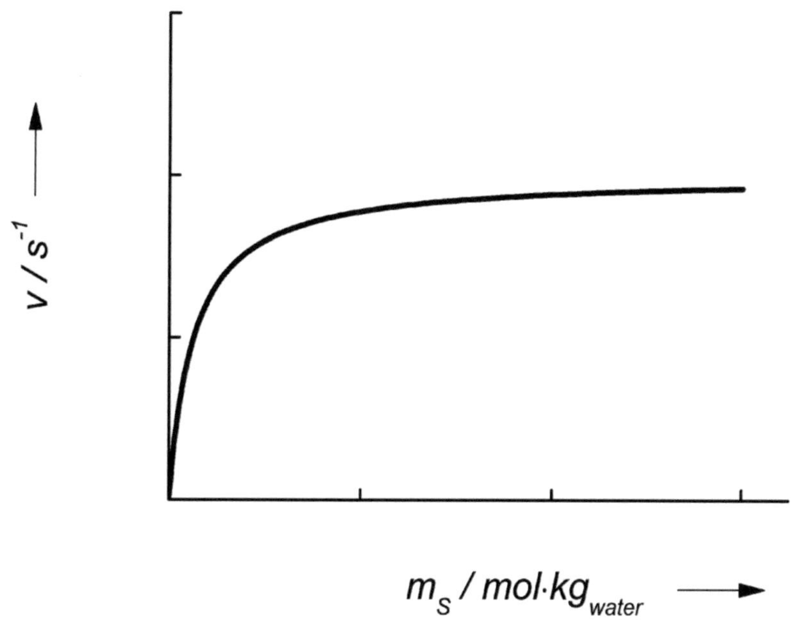

Figure 1. Qualitative Michaelis-Menten plot of the reaction rate v plotted over the substrate molality m_S according to Eq. (2).

As can be seen from Eq. (2) and **Figure 1**, the reaction rate follows a hyperbolic curve over increasing substrate concentrations. Further, K_M^{obs} defines the shape of the curve as it is the concentration of substrate at which the reaction velocity becomes half of its maximal value 0.5 v_{max}. Based on this, the importance of K_M^{obs} becomes obvious. If the value of K_M^{obs} is low compared to m_S required to reach $v = v_{max}$, it can be deduced that the cellular concentration of the substrates will also be close to K_M^{obs} as a significant increase in m_S (e.g., fivefold) will never increase v more than by a factor of 2 [9]. Thus, the knowledge of K_M^{obs} is of high importance for biology and for technical applications of enzyme-catalysed reactions.

Unfortunately, the majority of enzyme-catalysed reactions are not one-substrate reactions; in such cases, the reaction scheme increases in complexity. However, it is often still possible to apply pseudo-one-substrate reaction conditions given that the molality of one substrate is much higher than the molality required to obtain v_{max}. These conditions are obtained if substrate simultaneously presents the reaction solvent, which is the case for hydrolysis reactions. A general scheme for a two-substrate reaction is given in Eq. (3).

$$E + S_1 \rightleftharpoons ES_1 + S_2 \rightleftharpoons ES_1S_2 \rightarrow \quad E + P_1 + P_2 \tag{3}$$

In Eq. (3), substrates are labelled as *S1* and *S2*; the reaction mechanism (ordered or random) shall not be discussed at this point. In this case, the Michaelis-Menten equation changes to Eq. (4), which contains the Michaelis constants for substrate 1 K_{MS1}^{obs} and substrate 2 K_{MS2}^{obs} as well as the inhibition constant K_{iS1}^{obs}, which defines the reaction mechanism [8, 9].

$$v = \frac{v_{max} \cdot m_{S1} \cdot m_{S2}}{K_{iS1}^{obs} \cdot K_{MS2}^{obs} + K_{MS1}^{obs} \cdot m_{S2} + K_{MS2}^{obs} \cdot m_{S1} + m_{S_1} \cdot m_{S2}} \tag{4}$$

In the case of a hydrolysis reaction taking place in water as reaction solvent, the molality of substrate 2 m_{S2} (water) is usually two to three orders of magnitude higher than the molality of substrate 1, which gets cleaved by the enzyme. Rearranging Eq. (4) leads back to the Michaelis-Menten equation under this assumption shown in Eqs. (5)–(7).

$$v = \frac{v_{max} \cdot m_{S1}}{\frac{\left(K_{iS1}^{obs} \cdot K_{MS2}^{obs}\right)}{m_{S2}} + K_{MS1}^{obs} + K_{MS2}^{obs} \cdot \frac{m_{S1}}{m_{S2}} + m_{S_1}} \tag{5}$$

$$m_{S1} \gg m_{S2} \text{ and } m_{S1} \gg K_{iS1}^{obs} \cdot K_{MS2}^{obs} \tag{6}$$

$$v = \frac{v_{max} \cdot m_{S1}}{K_{M1}^{obs} + m_{S1}} \tag{7}$$

To be able to compare reactions from different research groups and further for different enzymes catalysing the same reaction, the Michaelis-Menten equation has to be normalised to the total enzyme concentration m_E according to Eq. (8).

$$\frac{d\frac{m_P}{m_E}}{dt} = \frac{dm_P}{dt} \cdot \frac{1}{m_E} = v' = \frac{k_{cat} \cdot m_S}{K_M^{obs} + m_S} \tag{8}$$

The determination of k_{cat} and K_M^{obs} is usually performed by measuring reaction rates for different substrate concentrations as shown in **Figure 1**. While this approach is common, it also poses a lot of difficulties and causes high uncertainties. The determination of k_{cat} requires that the solubility of the substrate has to be higher than the molality m_S that is required for reaching k_{cat}. Lowering the enzyme concentration and thus the required molality m_S often causes diffusion limitations that might lead to highly uncertain kinetic constants. Another possible issue is contrary, as a reaction might require high concentrations of an expensive substrate to determine k_{cat}. To overcome these possible limitations, the Lineweaver-Burk equation is commonly applied for the determination of the kinetic constants [19].

$$\underbrace{\frac{1}{v'}}_{y} = \underbrace{\frac{K_M^{obs}}{k_{cat}}}_{m} \cdot \underbrace{\frac{1}{m_S}}_{x} + \underbrace{\frac{1}{k_{cat}}}_{b} \tag{9}$$

Through this linearization, a plot of $(v')^{-1}$ over m_S^{-1} yields the kinetic constants: The slope (Sl) and the ordinate (Or) of the obtained linear fit can be used to determine K_M^{obs} and k_{cat} as given in Eqs. (10) and (11).

$$Sl = \frac{K_M^{obs}}{k_{cat}} \tag{10}$$

$$Or = \frac{1}{k_{cat}} \tag{11}$$

2.1.2. Activity-based approach

As presented in Section 2.1.1, the Michaelis constant is determined based on the molality m_S. If co-solvents are added to the neat reaction mixtures, the experimental procedure has to be performed also for the changed conditions. From the perspective of process design, this poses a huge cost-intensive and time-consuming approach towards finding suitable co-solvents for the desired application.

To be able to predict co-solvent influences on pseudo-one-substrate reactions, a thermodynamic co-solvent-independent Michaelis constant, further referred to as K_M^a, has to be determined. K_M^a is a constant value, which does not depend on co-solvent given that the co-solvent does not disturb the reaction mechanism (e.g., co-solvent acts as inhibitor or activator) and that the co-solvent has no denaturing effect on the enzyme. K_M^a can be determined under neat (co-solvent-free) conditions by replacing the molality m_S in Eq. (9) with thermodynamic activities of the substrate a_S. The latter are accessible by multiplying the concentration of a substrate by the respective concentration-based activity coefficient (molality-based γ_S^m, mole-fraction-based γ_S^x or molarity-based γ_S^c) as shown in Eq. (12) [20–22]:

$$a_S[-] = x_S \cdot \gamma_S^x = m_S \cdot \gamma_S^m = c_S \cdot \gamma_S^c \tag{12}$$

In the following, molality-based $\gamma_S{}^m$ will be used to analyse the data. Replacing molalities in Eq. (9) with activities leads to an activity-based Lineweaver-Burk equation:

$$\frac{1}{\nu'} = \frac{K_M^a}{k_{cat}} \cdot \frac{1}{a_S} + \frac{1}{k_{cat}} \tag{13}$$

To determine K_M^a, the experimental $(\nu')^{-1}$ values, which were determined for different m_S^{-1} values, are further plotted over the reciprocal substrate activity a_S^{-1}. It is noteworthy that this procedure does not change the value of k_{cat} [7, 23]. Under the assumption that the addition of a co-solvent only changes non-covalent interactions between the substrate and the other components in the reaction mixture, K_M^a is assumed to be a constant value. That is, any observed change in K_M^{obs} is directly reflected in $\gamma_S{}^{\mathrm{m}}$. With this knowledge, a prediction of K_M^{obs} under co-solvent influence becomes possible. For this, two hypothetical molalities of the substrate m_S^{hyp} are chosen randomly. Afterwards, activity coefficients of the substrate in the co-solvent system are predicted and the respective activities are calculated. Further, a random value of k_{cat} (e.g., value of the neat reaction) is chosen. Note that k_{cat} is a factor that cancels out during the linearization to determine K_M^{pre} (see Eq. (13)). In the next step, the predicted activities together with K_M^a and k_{cat} are used to predict $(\nu')^{-1}$ according to Eq. (13). Predicted $(\nu')^{-1}$ values are afterwards plotted over the chosen reciprocal molalities $\left(m_S^{hyp}\right)^{-1}$. In the final step, the predicted concentration-based Michaelis constant K_M^{pre} is determined according to Eq. (9). The process to determine K_M^{pre} is illustrated in **Scheme 1**.

As can be seen, the major aspect for the prediction of the Michaelis constants is the ability to predict the substrate activity coefficients. For this, a physically sound model, namely the electrolyte perturbed-chain statistical associating fluid theory (ePC-SAFT) was used in this work. This model has already been applied successfully to complex mixtures containing low-soluble molecules [24], PEG and salts [25] and electrolytes [20] while also being applied

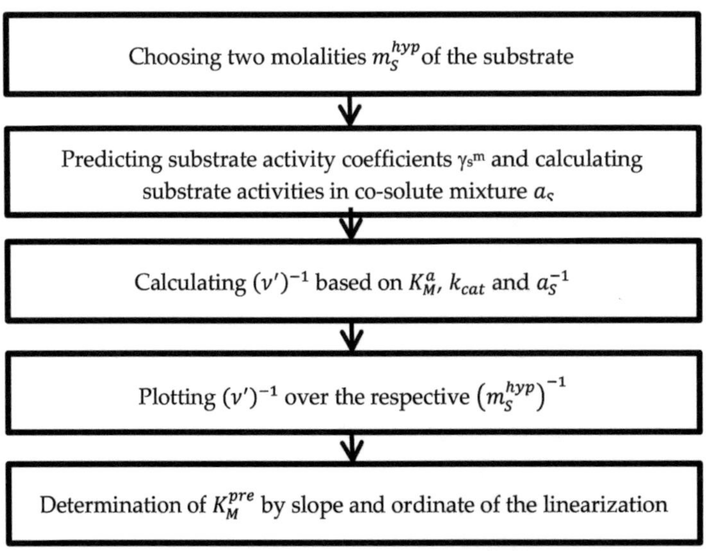

Scheme 1. Steps for the prediction of the concentration-based Michaelis constant K_M^{pre} under the influence of co-solvents. Predictions are based on the determined activity-based Michaelis constant K_M^a.

simultaneously to mixtures with up to eight components [4], and thus, it provides a reliable model basis for this work.

The ePC-SAFT equation of state is based on PC-SAFT, developed and proposed by Gross and Sadowski [26] and extended for electrolyte systems by Cameretti et al. [18] ePC-SAFT provides an expression for the residual Helmholtz energy a^{res} calculated from different contributions as shown in Eq. (14):

$$a^{res} = a^{hc} + a^{disp} + a^{assoc} + a^{ion} \tag{14}$$

In Eq. (14), the reference system is seen as a chain of hard spheres, which is represented by the contribution a^{hc}. The perturbations to this hard-chain reference system are accounted for in ePC-SAFT by the molecular dispersive interactions, characterised by the Van der Waals energy incorporated in a^{disp} and by the associative hydrogen bonding forces represented in a^{assoc}. As an addition for electrolyte systems, the Coulomb interactions based on the Debye-Hückel equation are expressed by a^{ion}. Based on a^{res}, fugacity coefficients φ can be calculated which allow determining the activity coefficients $\gamma_S{}^x$ using Eq. (15).

$$\gamma_S^x = \frac{\varphi_i\left(T, p, \vec{x}\right)}{\varphi_{0i}(T, p, x_i = 1)} \tag{15}$$

In Eq. (15), 0i denotes the pure component, which is the reference state at the same temperature T and pressure p as the actual solution of the composition $\vec{x}$. This means that activity coefficients can be estimated independent of the amount of components, temperature and pressure of the solution regarded. Eq. (15) is finally used with Eq. (12) to obtain the molality-based $\gamma_S{}^m$.

2.2. Kinetic assays

In this work, a pseudo-one-substrate reaction is presented using the hydrolysis of SPNA catalysed by the enzyme α-CT. The reaction mechanism is given in **Scheme 2**.

The kinetic measurements have been discussed already in [7] and are briefly summarised here. Lyophilized powder of α-CT was used as catalyst, and trimethylamine N-oxide (TMAO) and urea were used as co-solvents. Measurements were carried out in Tris-HCl buffer (100 mmol/ kg_{water} tris(hydroxymethyl)aminomethane, pH 8.0). The kinetic measurements of the neat and co-solvent reaction mixtures of the SPNA hydrolysis reactions were performed in a stopped-flow system (HPSF-56 of Hi-Tech Scientific) [27, 28]. In a first step, the substrate stock solution

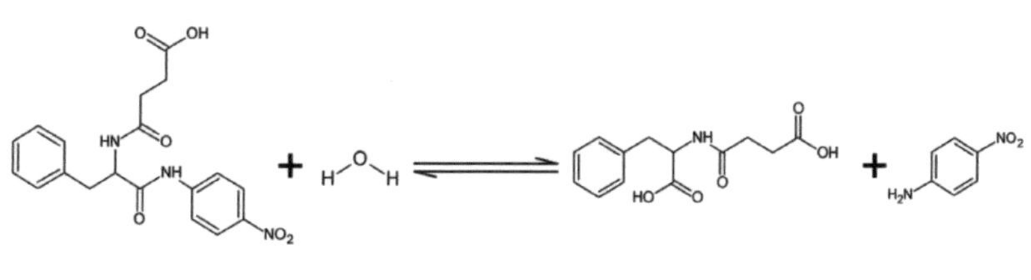

Scheme 2. Reaction scheme for the hydrolysis of SPNA catalysed by α-chymotrypsin. Products of the hydrolysis reaction are N-(3-carboxypropanoyl)phenylalanine and p-nitroaniline, respectively.

containing SPNA and the respective co-solvent in a 100 mol/kg$_{water}$ Tris-HCl buffer at pH 8 and the enzyme stock solution containing the respective co-solvent in a 100 mol/kg$_{water}$ Tris-HCl buffer at pH 8 were prepared and loaded for injection in the measurement cell. After simultaneous injection, the measurement cell was constantly monitored for the extinction at 410 nm wavelength, allowing the determination of the time-dependent change in the 4-NA concentration. The pH values of the stock solutions were measured directly before the start of the reaction to ensure no pH effect on K_M^{pre}; a pH electrode was used from Mettler Toledo with an uncertainty of ±0.01. The measured systems presented in this work are given in **Table 1**.

2.3. Results and discussion

In a first step, the concentration-based Michaelis constant K_M^{obs} was determined under neat conditions. The respective Lineweaver-Burk plot is given in **Figure 2**.

As can be seen from **Figure 2**, a linear relation between the reciprocal molality of the substrate m_{SPNA}^{-1} and the reciprocal normalised reaction rate $(\nu')^{-1}$ can be observed. This relation allowed the determination of K_M^{obs}; as a result, a value of 1.76 ± 0.12 mmol/kg$_{water}$ [7] was obtained for the SPNA hydrolysis. The activity-based Michaelis constant K_M^a was then obtained with activity coefficients of SPNA, which were predicted for each substrate molality with ePC-SAFT. The pure component and binary interaction parameters used for the ePC-SAFT prediction are listed in **Tables 2** and **3**.

Note that in a first step, mole-fraction-based activity coefficients were obtained with ePC-SAFT. Eq. (12) was used to convert these into molality-based activity coefficients; these were used throughout this work. In the next step, a plot of the determined $(\nu')^{-1}$ over the predicted reciprocal a_{SPNA}^{-1} was created. Based on this plot, K_M^a was determined in analogy to the determination of K_M^{obs} as shown in **Figure 2**, resulting in a value of $K_M^a = 0.0686$. This value was used as input value for the prediction of co-solvent influence on K_M^{obs} according to **Figure 2**. The comparison between this prediction and the experimental K_M^{obs} values is shown in **Figure 3** and **Table 4**.

Co-solvent	$m_{co-solvent}$ **(mol/kg$_{water}$)**	m_{SPNA} **(mmol/kg$_{water}$)**
Neat	—	0.125–1
TMAO	0.5	0.125–1
Urea	1	0.250–1
DMSO	2.1	0.250–1
DMSO	4.2	0.125–1

Enzyme concentration was 8 μmol/kg$_{water}$ in all kinetic assays.

Table 1. Overview of the measured systems to determine concentration-based Michaelis constants K_M^{obs}, adapted from [7], including the co-solvent and its concentration and the initial SPNA concentration range regarded.

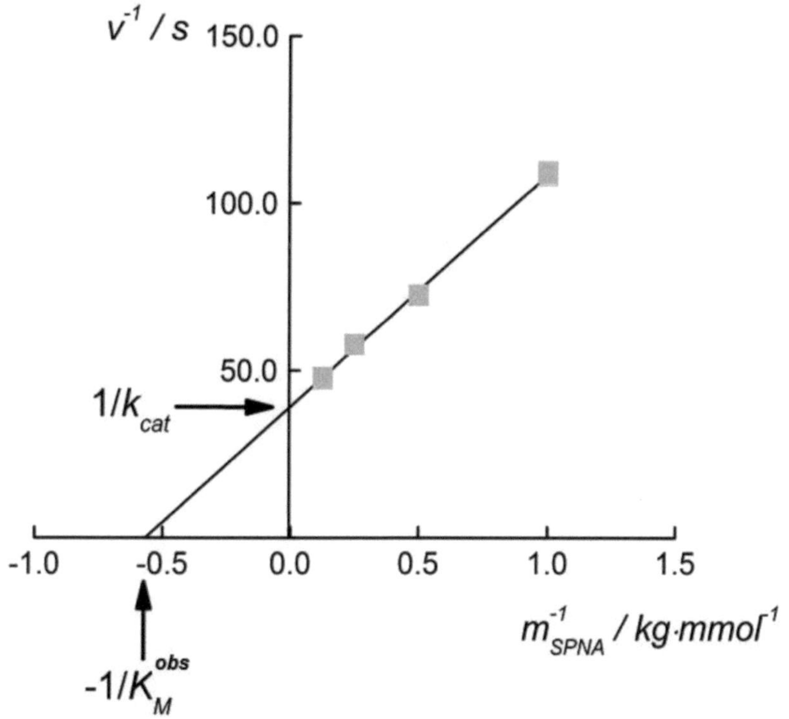

Figure 2. Lineweaver-Burk plot for the determination of the concentration-based Michaelis constant of SPNA K_M^{obs} at T = 25°C, p = 1 bar and pH = 8 in Tris-HCl buffer [7]. The plot shows experimental data points of the neat measurements (squares) which are obtained from the inverse turnover frequency $(\nu')^{-1}$ over the inverse substrate molality of SPNA m_{SPNA}^{-1}. K_M^{obs} was obtained by linear regression of the experimental data and extrapolation to the abscissa as shown.

Component	m_i (–)	σ_i (Å)	$\frac{u_i}{k_B}$ (K)	N_i^{assoc}	$\frac{\epsilon^{A_iB_j}}{k_B}$ (K)	$\kappa^{A_iB_i}$ (–)
Water [29]	1.204	[A]	353.95	1:1	2425.7	0.0451
DMSO [29]	2.922	3.28	355.69	1:1	0	0.0451
Urea [29]	4.242	2.45	368.23	1:1	3068.7	0.0010
TMAO [30]	8.93	2.25	245.44	1:1	0	0.0451
SPNA [7]	13.500	4.00	249.95	2:2	4351.0	0.0090

[A] $\sigma_i = 2.7927 + 10.11 \cdot exp(-0.01775 \cdot T) - 1.417 \cdot exp(-0.01146 \cdot T)$, T in Kelvin.

Table 2. ePC-SAFT pure-component parameters from [7, 29, 30].

As can be seen in **Figure 3** and **Table 4**, an accurate prediction of the co-solvent-induced changes in K_M^{obs} is possible. For both DMSO concentrations, predictions are even quantitatively correct within the experimental uncertainties. This is of special importance for the hydrolysis reaction under investigation since DMSO has the strongest impact on K_M^{obs}. The big advantage of K_M^a over K_M^{obs} is that it is a constant value independent of the co-solvent. This fact further

Mixture	k_{ij} (–)
Water-DMSO [30]	−0.065
Water-urea [30]	−0.044
Water-TMAO [30]	−0.149
Water-SPNA [7]	−0.132
DMSO-SPNA [7]	−0.117
Urea-SPNA [7]	−0.203
TMAO-SPNA [7]	−0.220

Table 3. ePC-SAFT binary interaction parameters [7, 29, 30].

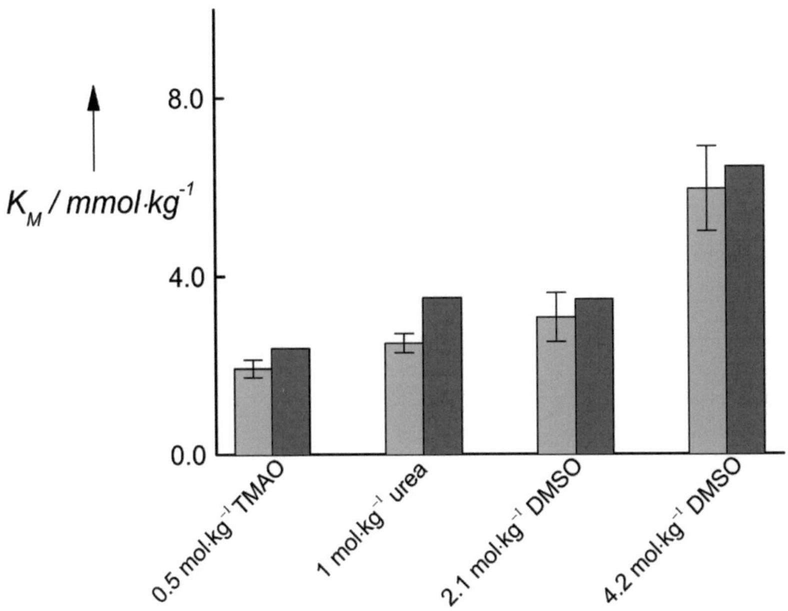

Figure 3. Comparison between experimental concentration-based Michaelis constants K_M^{obs} (light grey bars) at T = 25°C, p = 1 bar and pH = 8 in Tris-HCl buffer and the predicted Michaelis constants K_M^{pre} (dark grey bars). For the predictions, a constant K_M^{a} value of 0.0686 was used and the activity coefficients were predicted with ePC-SAFT based on the parameters from **Tables 2** and **3**. Reprinted from [7].

allows predicting K_M^{obs} without the need for additional experimental data. This proves the validity of the proposed modelling approach for pseudo-one-substrate reactions and validates the assumption that co-solvent-substrate interactions are responsible for the dependence of K_M^{obs} on co-solvents. Thus, this indirectly disproves that enzyme-co-solvent effects are responsible for such changes of K_M^{obs}. However, as enzyme-catalysed reactions are mostly multi-substrate reactions with two substrates of low concentrations, the following section presents the transfer of the gained insight and methods to two-substrate reactions.

http://dx.doi.org/10.5772/intechopen.80235

Co-solvent	$m_{co\text{-}solvent}$ (mol/kg$_{water}$)	K_M^{obs} $(mmol/kg_{water})$	K_M^{pre} $(mmol/kg_{water})$
TMAO	0.5	1.93 ± 0.19	2.38
Urea	1	2.50 ± 0.21	3.51
DMSO	2.8	3.08 ± 0.54	3.48
DMSO	4.2	5.96 ± 0.95	6.45

K_M^{pre} were predicted using K_M^a determined from experimental $K_M^{obs} = 1.76\ mmol/kg_{water}$ of the neat reaction [7].

Table 4. Comparison between the experimental K_M^{obs} with the respective predicted values K_M^{pre} under the influence of the co-solvents TMAO, urea or DMSO at $T = 25°C$, $p = 1$ bar and pH = 8 in Tris-HCl buffer.

3. Two-substrate reactions

3.1. Theoretical background

3.1.1. Concentration-based approach

As presented in Section 2.1.1 ('pseudo') one-substrate reactions occur seldom in enzyme catalysis. Enzyme catalysis often requires co-substrate that is present in a limiting concentration (e.g., NADH, ATP, GTP). A two-substrate reaction can be described by Eq. (16), which cannot be simplified further:

$$v' = \frac{k_{cat} \cdot m_{S1} \cdot m_{S2}}{K_{iS1}^{obs} \cdot K_{MS2}^{obs} + K_{MS1}^{obs} \cdot m_{S2} + K_{MS2}^{obs} \cdot m_{S1} + m_{S_1} \cdot m_{S2}} \tag{16}$$

Two-substrate reactions can have a specific binding order attached to them. To account for this, the inhibition constant of S1 K_{iS1}^{obs} based on the Haldane relation was accounted for in this work; if K_{iS1}^{obs} is lower than $K_{M,S1}^{obs}$, an ordered mechanism is present in which S1 has to bind first [8, 31]. The Lineweaver-Burk linearization of Eq. (16) leads to Eq. (17):

$$\underbrace{\frac{1}{v'}}_{y} = \cdot \underbrace{\left(\frac{K_{iS1}^{obs} \cdot K_{M,S2}^{obs}}{k_{cat} \cdot m_{S2}} + \frac{K_{MS1}^{obs}}{k_{cat}}\right)}_{m} \cdot \underbrace{\frac{1}{m_{S1}}}_{x} + \underbrace{\left(\frac{K_{MS2}^{obs}}{k_{cat} \cdot m_{S2}} + \frac{1}{k_{cat}}\right)}_{b} \tag{17}$$

Note that Eq. (17) does not show any direct relation between the Michaelis constants and the ordinate, slope or the abscissa of the linearization. In the case of two-substrate reactions, a two-step linearization process is suggested. For this, the molality of one of the substrates, in this case m_{S2}, is chosen to be at least 50 times higher than m_{S1}. Under this assumption, a so-called *primary plot* can be created. For this, different levels of m_{S2} that are regarded to be constant over the short reaction time are chosen for varying m_{S1}; then, a family of straight lines are obtained as shown exemplarily in **Figure 4**.

Each of the straight lines in **Figure 4** has its own slope (Sl^{prim}) and ordinate (Or^{prim}); both, Sl^{prim} and Or^{prim} are a function of m_{S2} as shown in Eqs. (18) and (19).

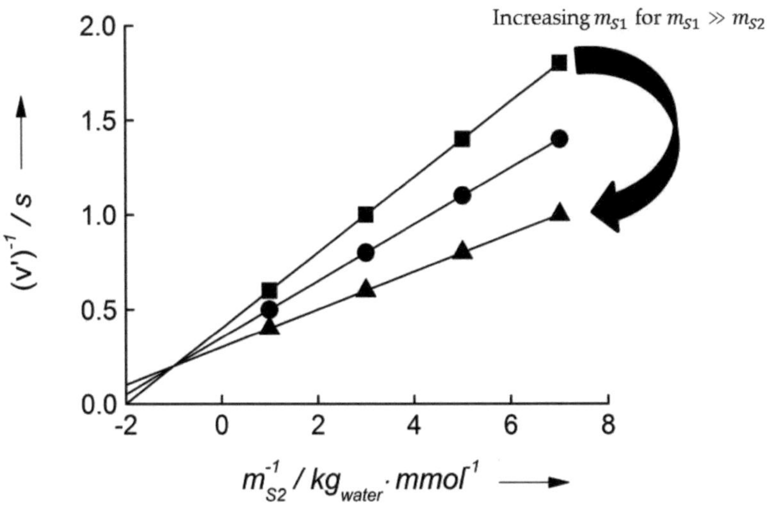

Figure 4. Exemplary *primary plot* for a two-substrate reaction obtained from plotting the inverse turnover frequency $(v')^{-1}$ over the inverse substrate molality of substrate 2 m_{S2}^{-1} for different pseudo-constant molalities of substrate 1 m_{S1}. Molalities m_{S1} increase in the order of $m_{S1,\,squares} > m_{S1,\,circles} > m_{S1,\,triangles}$.

$$Sl^{prim} = \frac{K_{iS1}^{c} \cdot K_{MS2}^{c}}{k_{cat}} \cdot \frac{1}{m_{S2}} + \frac{K_{MS1}^{c}}{k_{cat}} \tag{18}$$

$$Or^{prim} = \frac{K_{MS2}^{c}}{k_{cat}} \cdot \frac{1}{m_{S2}} + \frac{1}{k_{cat}} \tag{19}$$

Eqs. (18) and (19) again show a linear correlation between Sl^{prim} and m_{S2}^{-1} as well as between Or^{prim} and m_{S2}^{-1}, respectively. This allows for another linearization step represented in two *secondary plots* in **Figure 5**.

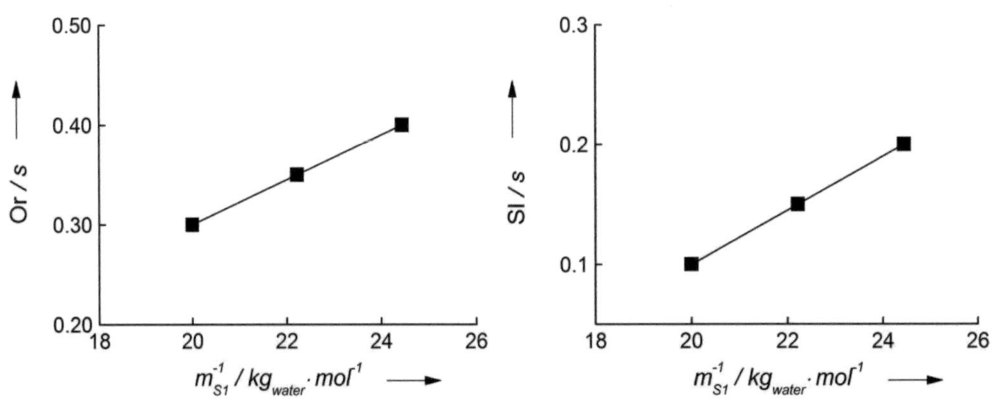

Figure 5. Exemplary *secondary plot* for Or^{prim} (left) and Sl^{prim} (right) over the reciprocal pseudo-constant molalities of m_{S1}^{-1} derived from the *primary plot* given in **Figure 4**.

Sl_{Sl}^{sec}, Or_{Sl}^{sec}, Sl_{Or}^{sec} and Or_{Or}^{sec} obtained from the *secondary plots* are defined according to Eqs. (20)–(23):

$$Sl_{Sl}^{sec} = \frac{K_{iS1}^{obs} \cdot K_{MS2}^{obs}}{k_{cat}} \tag{20}$$

$$Or_{Sl}^{sec} = \frac{K_{MS1}^{obs}}{k_{cat}} \tag{21}$$

$$Sl_{Or}^{sec} = \frac{K_{MS2}^{obs}}{k_{cat}} \tag{22}$$

$$Or_{Or}^{sec} = \frac{1}{k_{cat}} \tag{23}$$

The relations shown in Eqs. (20)–(23) are finally used to determine K_{MS1}^{obs}, K_{MS2}^{obs}, k_{cat} and K_{iS1}^{obs}.

3.1.2. Activity-based approach

The determination of activity-based Michaelis constants $K_{M,S1}^{a}$ and $K_{M,S2}^{a}$ for two-substrate reactions is analogous to pseudo-one-substrate reactions. As for the pseudo-one-substrate reaction, molalities in Eq. (17) are replaced with activities as shown in Eq. (24):

$$\underbrace{\frac{1}{v'}}_{y} = \cdot \underbrace{\left(\frac{K_{iS1}^{a} \cdot K_{M,S2}^{a}}{k_{cat} \cdot a_{S2}} + \frac{K_{MS1}^{a}}{k_{cat}}\right)}_{m} \cdot \underbrace{\frac{1}{a_{S1}}}_{x} + \underbrace{\left(\frac{K_{MS2}^{a}}{k_{cat} \cdot a_{S2}} + \frac{1}{k_{cat}}\right)}_{b} \tag{24}$$

From Eq. (24), a *primary plot* is created as described in Section 3.1.1 in which $(v')^{-1}$ is plotted over a_{S1}^{-1}. Afterwards, the two *secondary plots* are created by plotting the Or^{prim} and Sl^{prim} of the *primary plot* over a_{S1}^{-1} to finally obtain the activity-based kinetic constants K_{iS1}^{a}, K_{MS1}^{a} and K_{MS2}^{a} as described for the concentration-based approach in Section 3.1.1.

Predictions for the co-solvent influence on K_{MS1}^{obs} and K_{MS2}^{obs} are performed in analogy to pseudo-one-substrate reactions: Two molalities of S1 m_{S1}^{hyp} for two molalities of S2 m_{S2}^{hyp} have to be chosen; then, the required activity coefficients have to be predicted in order to create a predicted *primary plot;* the *secondary plots* are then constructed by plotting Or^{prim} and Sl^{prim} over the chosen reciprocal molalities $\left(m_{S2}^{hyp}\right)^{-1}$. In a final step, the predicted Michaelis constants K_{MS1}^{pre} and K_{MS2}^{pre} are obtained from the *secondary plots*. The prediction process is illustrated in **Scheme 3**.

3.2. Materials and methods

In this work, the reduction of acetophenone by two different enzymes, ADH 270 and ADH 200, was investigated as model reaction for a two-substrate reaction. The reaction scheme is given in **Scheme 4**. Kinetic data for the ADH 270 were taken from [23].

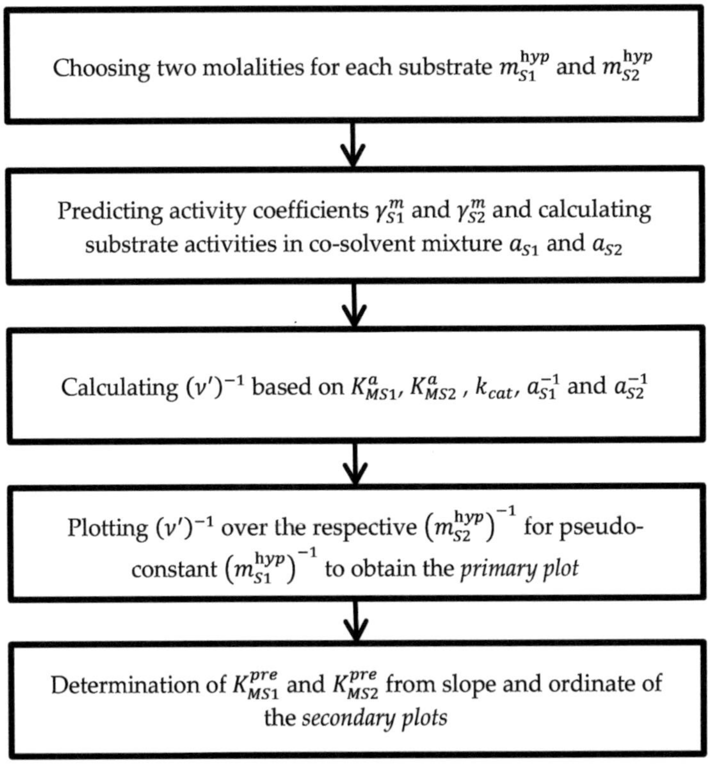

Scheme 3. Steps for the prediction of the concentration-based Michaelis constants K_{MS1}^{pre} and K_{MS2}^{pre} under the influence of co-solvents. Predictions are based on the determined activity-based Michaelis constants K_{MS1}^{a} and K_{MS2}^{a}.

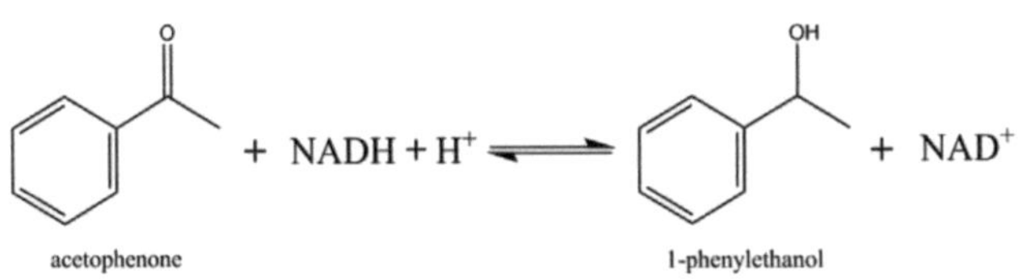

Scheme 4. Reaction scheme for the reduction of acetophenone to 1-phenylethanol with the co-substrate nicotinamide adenine dinucleotide in its protonated form (NADH+H+) and its deprotonated form (NAD+) catalysed by two different genetically modified alcohol dehydrogenases recombinant from *E. coli* (evo-1.1.270; evo-1.1.200).

3.2.1. Chemicals

2-[4-(2-hydroxyethyl)piperazin-1-yl]ethanesulfonic acid (HEPES) and polyethylene glycol 6000 (PEG 6000) were purchased from VWR. Acetophenone (ACP) and NADH were purchased from Sigma Aldrich. Sodium hydroxide was purchased from Bernd Kraft GmbH. The genetically modified enzyme alcohol dehydrogenase 200 (evo-1.1.20) expressed recombinant in *E. coli* was purchased from Evoxx. All chemicals were used without further purification, and all samples were prepared using Millipore water from the Milli-Q provided by Merck Millipore as stated in the chemical provenance (**Table 5**). Kinetic results using the genetically modified enzymes alcohol dehydrogenase 270 (evo-1.1.270) were taken from [23].

http://dx.doi.org/10.5772/intechopen.80235

Compound	Purity	CAS	Supplier
2-[4-(2-hydroxyethyl)piperazin-1-yl]ethanesulfonic acid (HEPES)	>99%	7365-459	VWR
Polyethylene glycol 6000	—	25322-68-3	VWR
Acetophenone (ACP)	>99%	98-86-2	S
NADH	>97%	606-68-8	S
Sodium hydroxide	>98%	1310-73-2	BK
Alcohol dehydrogenase 200 (evo-1.1.200)	30%	evo-1.1.200	E

S = Sigma Aldrich Chemie GmbH; VWR = VWR International GmbH; BK = Bernd Kraft GmbH; E = Evoxx technologies GmbH.

Table 5. Chemical provenance table for the components measured in this work.

3.2.2. Kinetic assays

Reactions were carried out in an HEPES buffer (0.1 mol/kg$_{water}$) at pH 7. The pH of the buffer and each sample was measured using a pH electrode from Mettler Toledo (uncertainty ±0.01) and adjusted with sodium hydroxide when required. For measurements of the co-solvent influence, 17 wt.% of PEG 6000 was added to the buffer. In the first step, the substrate solutions of ACP were prepared in equal number to the different NADH concentrations measured. Buffer was added to 5 ml Eppendorf cups, and ACP was added gravimetrically afterwards using the XS analytical balance provided by Mettler Toledo (uncertainty ±0.01 mg). Eppendorf cups were filled to the maximum capacity in order to decrease losses of ACP to the vapour phase. The ACP stock solutions were preheated in an Eppendorf ThermoMixer C at 25°C. ACP concentrations of the neat reaction were 20, 30 and 40 mmol/kg$_{water}$ and 60, 80 and 100 mmol/kg$_{water}$ for the PEG 6000 measurements, respectively. NADH was added gravimetrically to the ACP solution after preheating. Each sample was prepared directly before measurements due to reported long-term instability of NADH in solution [32]. NADH concentrations were chosen to be 0.15, 0.2, 0.25, 0.3, 0.35 and 0.4 mmol/kg$_{water}$. The enzyme stock solution was prepared by gravimetrically adding 1 wt.% of enzyme to 2 ml of buffer with direct storage on ice for the period of all measurements to ensure enzyme stability and activity. To initiate the kinetic measurements, 20 mg of the enzyme solution was transferred into a quartz cuvette SUPRASIL TYP 114-QS from Helma Analytics which was preheated to 25°C while being placed in an Eppendorf Biospectrometer. After addition of 1 g of the substrate solution containing ACP and NADH, the measurement of the extinction over time at 340 nm wavelength was initiated.

3.3. Results and discussion

3.3.1. ADH 270

In a first step, the *primary plot* for the ACP reduction catalysed by ADH 270 was determined under neat conditions. For this, $(\nu')^{-1}$ is plotted over m_{NADH}^{-1} for pseudo-constant m_{ACP} levels of 20, 30 and 40 mmol/kg$_{water}$ in **Figure 6**.

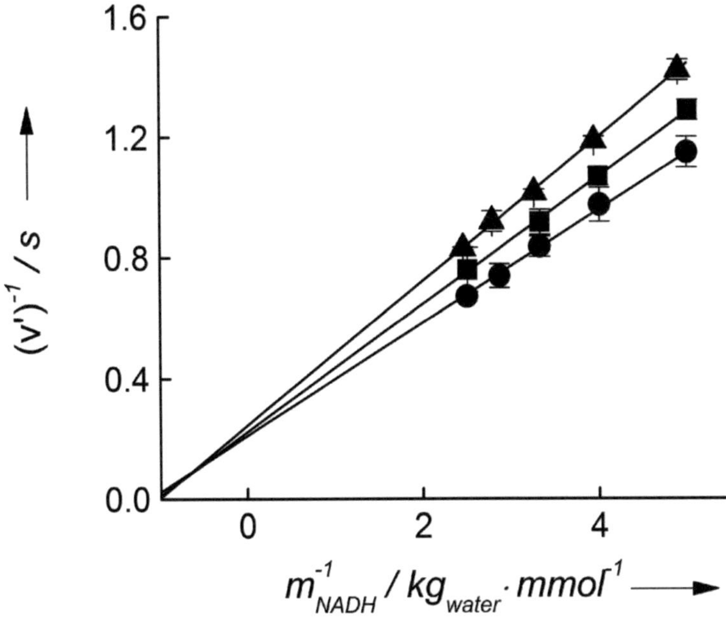

Figure 6. *Primary plot* based on Eq. (17) for the ACP reduction catalysed by ADH 270 under neat conditions at T = 25°C, p = 1 bar and pH = 7 in HEPES buffer [23]. The reciprocal turnover frequency normalised to the total enzyme concentration $(\nu')^{-1}$ is plotted over the reciprocal initial molality of NADH m_{NADH}^{-1} for ACP molalities of 20 (triangles), 30 (squares) and 40 mmol/kg$_{\text{water}}$ (circles). Lines represent the respective fit lines required for further data analysis.

A linear correlation of $(\nu')^{-1}$ over m_{NADH}^{-1} can be observed from **Figure 6**. As described in Section 3.1.1, this correlation is used for the creation of the *secondary plots* shown in **Figure 7**.

Figure 7 shows the required linear correlation between Or^{prim} and Sl^{prim} over m_{ACP}^{-1} required for the determination of the Michaelis constants $K_{M,NADH}^{obs}$ and $K_{M,ACP}^{obs}$ according to Eqs. (20)–(23). Applying the relations given in Eqs. (20)–(23), the Michaelis constant of NADH

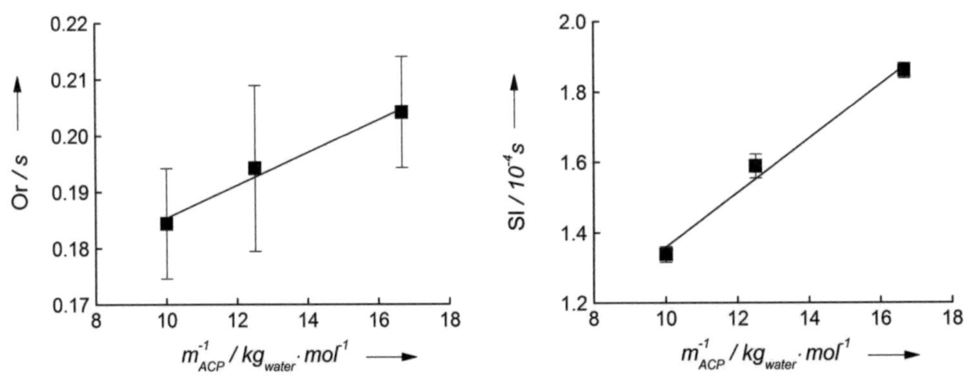

Figure 7. *Secondary plots* based on Eqs. (18) and (19) for the ACP reduction catalysed by ADH 270 under neat conditions at T = 25°C, p = 1 bar and pH = 7 in HEPES buffer [23]. Left: Ordinates Or^{prim} of the fit lines resulting from the primary plot are plotted over the reciprocal initial molality of ACP m_{ACP}^{-1}. Right: Slopes Sl^{prim} of the fit lines resulting from the primary plot are plotted over the reciprocal initial molality of ACP m_{ACP}^{-1}. Lines represent the respective fit lines required for further data analysis.

$K_{M,NADH}^{obs} = 0.37 \pm 0.09\ mmol/kg_{water}$ and of ACP $K_{M,ACP}^{obs} = 18.56 \pm 3.23\ mmol/kg_{water}$ were determined [23]. Afterwards, activity coefficients for NADH and ACP were predicted for the respective molalities; these ePC-SAFT predictions are based on the pure-component and binary interaction parameters listed in **Tables 6** and **7**. Based on the activities of NADH a_{NADH} and ACP a_{ACP}, the respective *primary* and *secondary plots* were created.

In analogy to the concentration-based approach, activity-based Michaelis constants were determined (Section 3.1.2) to be $K_{M,NADH}^{a} = 5.649 \cdot 10^{-8}$ and $K_{M,ACP}^{a} = 0.640$. As can be seen, activity-based Michaelis constants can be completely different from their concentration-based pendants; they even do not have any unit due to the definition of the activity as shown in Eq. (12). In the next step, a prediction of the co-solvent influence of 17 wt.% of PEG 6000 on $K_{M,ACP}^{obs}$ and $K_{M,NADH}^{obs}$ was performed as described in Section 3.2.1. These predictions were compared to experimental results given in **Figure 8** and **Table 8**.

As can be seen from **Figure 8** and **Table 8**, predictions of the co-solvent influence of 17 wt.% of PEG on $K_{M,NADH}^{obs}$ and $K_{M,ACP}^{obs}$ of the ACP reduction catalysed by ADH 270 are in very good agreement with experimental data. Upon addition of 17 wt.% PEG 6000, $K_{M,NADH}^{obs}$ decreased by a factor of two, while $K_{M,ACP}^{obs}$ increased by a factor of 2.5. Both trends were predicted

Component	m_i (−)	σ_i (Å)	$\frac{u_i}{k_B}$ (K)	N_i^{assoc}	$\frac{\varepsilon^{A_iB_j}}{k_B}$ (K)	$\kappa^{A_iB_i}$ (−)
Water [29]	1.204	[A]	353.95	1:1	2425.7	0.0451
ACP [4]	3.40	3.65	322.00	1:1	0	0.0451
NADH [33]	27.27	2.21	260.72	8:8	358.2	0.0001
PEG [25]	$M_{PEG} \cdot 0.05$	2.90	204.60	4:4	1799.8	0.020
Na^+ [34]	1	2.82	230	—	—	—
OH^- [34]	1	2.02	650.00	—	—	—

[A] $\sigma_i = 2.7927 + 10.11 \cdot exp(-0.01775 \cdot T) - 1.417 \cdot \exp(0.01146 \cdot T)$.

Table 6. ePC-SAFT pure-component parameters.

Binary pair	k_{ij} (−)
Water-ACP [4]	0.0330
Water-NADH [33]	−0.0585
Water-PEG [25]	[A]
Water-Na^+ [34]	[B]
Water-OH^- [34]	−0.25
Na^+-OH^- [34]	0.649

[A] $k_{ij}(T) = -0.135 + 0.0023439 \cdot (T[K] - 298.15)$.
[B] $k_{ij}(T) = 0.00045485 - 0.007981 \cdot (T[K] - 298.15)$.

Table 7. ePC-SAFT binary interaction parameters k_{ij}.

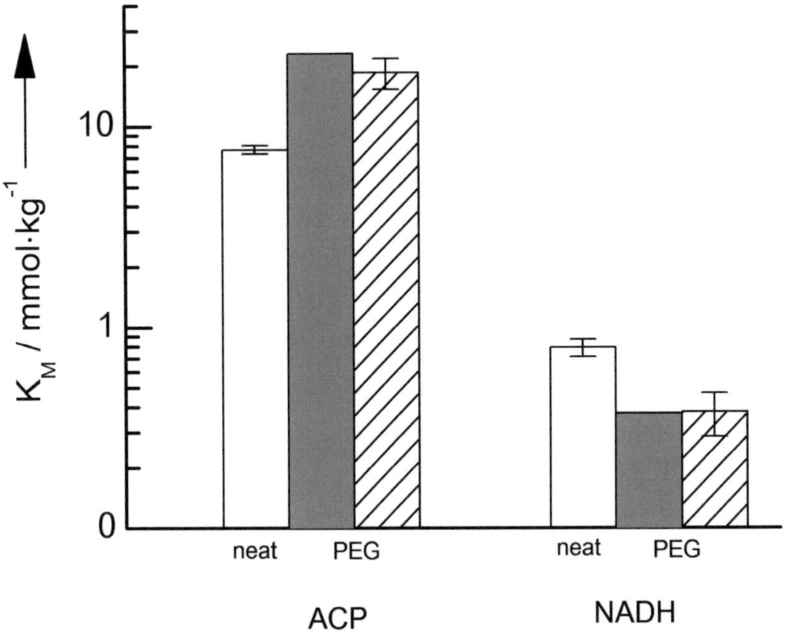

Figure 8. Comparison between the experimentally measured Michaelis constants K_M^{obs} of ACP and NADH under neat reaction conditions (white bars) and under the influence of 17 wt.% PEG 6000 (striped bars) for the reduction of ACP catalysed by ADH 270 at $T = 25°C$, $p = 1$ bar and pH = 7. The grey bars present the prediction of the respective K_M^{pre} based on the determined activity-based K_M^a from the experimental neat data [23]. Required activity coefficients were calculated with ePC-SAFT based on the parameters from **Tables 6** and 7.

	$K_{M,NADH}^{obs}\left[\frac{\text{mmol}}{\text{kg}_{\text{water}}}\right]$	$K_{M,NADH}^{pre}\left[\frac{\text{mmol}}{\text{kg}_{\text{water}}}\right]$	$K_{M,ACP}^{obs}\left[\frac{\text{mmol}}{\text{kg}_{\text{water}}}\right]$	$K_{M,ACP}^{pre}\left[\frac{\text{mmol}}{\text{kg}_{\text{water}}}\right]$
Neat	0.79 ± 0.08	–	7.67 ± 0.37	–
17 wt.% PEG 6000	0.377 ± 0.09	0.372	18.56 ± 3.23	23.00

Activity coefficients required for the prediction were calculated with ePC-SAFT based on the parameters from **Tables 6** and 7.

Table 8. Overview of the Michaelis constants under neat reaction conditions and the comparison between predicted K_M^{pre} and experimentally determined Michaelis constants K_M^{obs} [23] for the reduction of ACP catalysed by ADH 270 at $T = 25°C$, $p = 1$ bar and pH = 7.

accurately with ePC-SAFT. This shows that the influence of PEG 6000 on the K_M^{obs} values is caused by non-covalent molecular interactions between the co-solvent and the substrates instead of co-solvent-enzyme interactions.

3.3.2. ADH 200 and comparison to ADH 270

To further validate this approach, the ACP reduction was also investigated with ADH 200 as catalyst. This step is important to support the hypothesis that co-solvent-substrate interactions determine the co-solvent influence on $K_{M,ACP}^{obs}$ and $K_{M,NADH}^{obs}$. However, it becomes obvious from **Table 9** that ADH 200 shows a completely different kinetic profile under neat conditions.

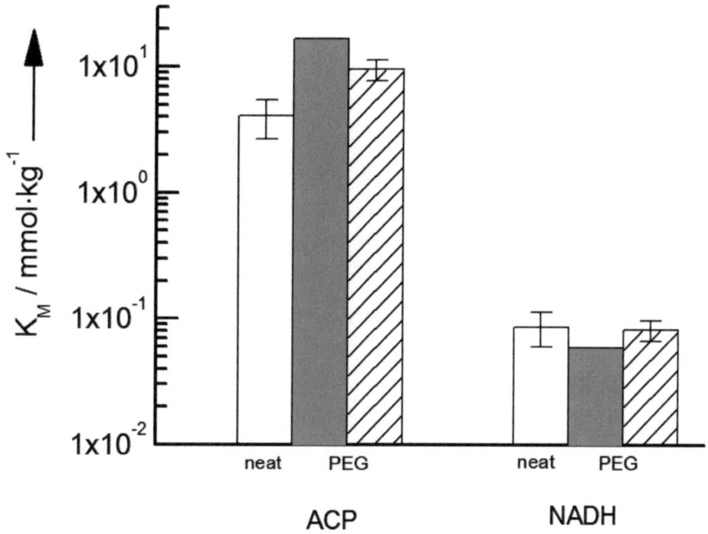

Figure 9. Comparison between the experimentally measured Michaelis constants K_M^{obs} for ACP and NADH from this work under neat reaction conditions (white bars) and under the influence of 17 wt.% PEG 6000 (striped bars) for the reduction of ACP catalysed by ADH 200 at $T = 25°C$, $p = 1$ bar and pH = 7. The grey bars present the predicted values for K_M^{pre} based on the determined activity-based K_M^a from the experimental neat data. Required activity coefficients were predicted with ePC-SAFT based on the parameters from **Tables 6** and **7**.

Enzyme	$K_{M,NADH}^{obs}\left[\frac{mmol}{kg_{water}}\right]$	$K_{M,NADH}^{a}[-]$	$K_{M,ACP}^{obs}\left[\frac{mmol}{kg_{water}}\right]$	$K_{M,ACP}^{a}[-]$
ADH 270 [23]	0.79 ± 0.08	$5.65 \cdot 10^{-8}$	7.67 ± 0.37	0.640
ADH 200 [this work]	0.086 ± 0.027	$1.16 \cdot 10^{-8}$	4.08 ± 1.03	0.749

Table 9. Comparison between the Michaelis constants of NADH and ACP for the reduction of ACP for neat reaction conditions at T=25 °C, p=1 bar and pH=7 in HEPES buffer. Two different enzymes were used as catalyst, ADH 270 and ADH 200. Activity coefficients required for the prediction of K_M^a were calculated with ePC-SAFT based on the parameters from **Table 6** and **Table 7**.

Table 9 shows that $K_{M,NADH}^{obs}$ ($K_{M,ACP}^{obs}$) using ADH 200 are 9 times (2 times) lower than $K_{M,NADH}^{obs}$ $\left(K_{M,ACP}^{obs}\right)$ using ADH 270 for identical conditions. Nevertheless, this is an expected behaviour. It can be further observed from **Table 9** that also the activity-based Michaelis constants $K_{M,NADH}^{a}$ and $K_{M,ACP}^{a}$ are different for ADH 200 and ADH 270 for the ACP reduction at same conditions. The prediction results of the influence of 17 wt.% PEG 6000 on K_M^{obs} of the reaction catalysed by ADH 200 are given in **Figure 9**.

Figure 9 shows that ePC-SAFT is able to predict the change of $K_{M,NADH}^{obs}$ and $K_{M,ACP}^{obs}$ under the influence of 17 wt.% of PEG 6000 for ADH 200 in good agreement with experimental data. The same ePC-SAFT parameters were used as for the prediction of the same reaction catalysed by ADH 270. This is a further validation of our approach as it shows that predictions are possible independent of the enzyme catalysing the reaction. For both enzymes, ADH 200 and ADH 270, the co-solvent influences on the substrate activities were the key for predicting the change in K_M^{obs}.

4. Conclusion

In this work, it was found that experimental Michaelis constants K_M^{obs} of a pseudo-one-substrate and a two-substrate reaction were strongly dependent on the co-solvent present in the reaction mixture. This co-solvent effect was assumed to be introduced by the thermodynamic non-ideality caused by molecular interactions. These are expressed as activity coefficients of the substrate(s), which were predicted by the equation of state ePC-SAFT. By accounting for the activity coefficients of the substrate(s), the concentration-based Michaelis constants K_M^{obs} were expressed as activity-based values K_M^a. This approach focused on investigating enzyme-independent interactions between co-solvent and the substrate(s) of the reaction; this has the advantage that K_M^a is a constant value independent of kind or concentration of co-solvent, while the experimentally observed K_M^{obs} values depend on co-solvent. The availability of K_M^a then allowed predicting co-solvent-induced changes in K_M^{obs} and therewith (1) proved the hypothesis that substrate-co-solvent interactions are responsible for changes of K_M^{obs} upon co-solvent addition and (2) enzyme-co-solvent interactions do not play a role for the observed changes in K_M^{obs}. Based on these findings, we could suggest that K_M^a should be considered instead of K_M^{obs} for investigations of enzyme-catalysed reactions in order to significantly reduce experimental effort and to gain new insight and understanding of the co-solvent-substrate-enzyme interactions present in more complex reaction mixtures, approaching *in cellulo* reaction conditions. We showed the feasibility of this by accurately predicting the influence of co-solvents (e.g., DMSO, urea or TMAO) on the Michaelis constants of a pseudo-one-substrate reaction as well as of a two-substrate ADH reaction. For the latter, the predictions were accurate for two different enzymes (ADH 200 and ADH 270) under investigation. This can be seen as another validation of the hypotheses (1) and (2).

Acknowledgements

AW and CH gratefully acknowledge the financial support of DAAD (project number 57340264) funded by the Federal Ministry of Education and Research (BMBF). Further, this work was supported by the Cluster of Excellence RESOLV (EXC 1069) funded by the Deutsche Forschungsgemeinschaft (DFG).

Conflict of interest

The authors declare no conflict of interest. Note that reference [23] is still under second review.

Author details

Anton Wangler, Mark Jonathan Bunse, Gabriele Sadowski and Christoph Held*

*Address all correspondence to: christoph.held@tu-dortmund.de

Department of Biochemical and Chemical Engineering, Laboratory of Thermodynamics, TU Dortmund University, Dortmund, Germany

References

[1] Straathof AJJ, Panke S, Schmid A. The production of fine chemicals by biotransformations. Current Opinion in Biotechnology. 2002;**13**(6):548-556

[2] Carrea G, Riva S. Properties and synthetic applications of enzymes in organic solvents. Angewandte Chemie International Edition. 2000;**39**(13):2226-2254

[3] Schulze B, Wubbolts MG. Biocatalysis for industrial production of fine chemicals. Current Opinion in Biotechnology. 1999;**10**(6):609-615

[4] Voges M et al. Measuring and predicting thermodynamic limitation of an alcohol dehydrogenase reaction. Industrial & Engineering Chemistry Research. 2017;**56**(19):5535-5546

[5] Voges M et al. Thermodynamics of the alanine aminotransferase reaction. Fluid Phase Equilibria. 2016;**422**:87-98

[6] Wangler A et al. Standard Gibbs energy of metabolic reactions: III the 3-phosphoglycerate kinase reaction. ACS Omega. 2018;**3**(2):1783-1790

[7] Wangler A et al. Co-solvent effects on reaction rate and reaction equilibrium of an enzymatic peptide hydrolysis. Physical Chemistry Chemical Physics. 2018;**20**(16):11317-11326

[8] Bisswanger H. Enzymkinetik: Theorie und Methoden. 3rd ed. Weinheim: Wiley-VCH; 2000

[9] Segel IH. Enzyme Kinetics: Behavior and Analysis of Rapid Equilibrium and Steady State Enzyme Systems. New York: Wiley; 1975

[10] Grosch J-H et al. Thermodynamic activity-based intrinsic enzyme kinetic sheds light on enzyme-solvent interactions. Biotechnology Progress. 2017;**33**(1):96-103

[11] Smith RR, Canady WJ. Solvation effects upon the thermodynamic substrate activity; correlation with the kinetics of enzyme catalyzed reactions. II. More complex interactions of alpha-chymotrypsin with dioxane and acetone which are also competitive inhibitors. Biophysical Chemistry. 1992;**43**(2):189-195

[12] Verma SK, Ghosh KK. Catalytic activity of enzyme in water/organic cosolvent mixtures for the hydrolysis of p-nitrophenyl acetate and p-nitrophenyl benzoate. Indian Journal of Chemistry. 2010;**49**:1041-1046

[13] Eremeev NL. Interaction of α-chymotrypsin with dimethyl sulfoxide: A change of substrate could "change" the interaction mechanism. Russian Journal of Bioorganic Chemistry. 2003;**29**(5):434-440

[14] Pleiss J. Thermodynamic activity-based interpretation of enzyme kinetics. Trends in Biotechnology. 2017;**35**(5):379-382

[15] Pleiss J. Thermodynamic activity-based progress curve analysis in enzyme kinetics. Trends in Biotechnology. 2018;**36**(3):234-238

[16] Chen CC, Song Y. Generalized electrolyte-NRTL model for mixed-solvent electrolyte systems. AICHE Journal. 2004;**50**(8):1928-1941

[17] Fredenslund A, Jones Russell L, Prausnitz John M. Group-contribution estimation of activity coefficients in nonideal liquid mixtures. AICHE Journal. 1975;**21**(6):1086-1099

[18] Cameretti LF, Sadowski G, Mollerup JM. Modeling of aqueous electrolyte solutions with perturbed-chain statistical associated fluid theory. Industrial & Engineering Chemistry Research. 2005;**44**(9):3355-3362

[19] Dowd JE, Riggs DS. A comparison of estimates of Michaelis-Menten kinetic constants from various linear transformations. Journal of Biological Chemistry. 1965;**240**(2):863-869

[20] Held C, Cameretti LF, Sadowski G. Modeling aqueous electrolyte solutions: Part 1. Fully dissociated electrolytes. Fluid Phase Equilibria. 2008;**270**(1–2):87-96

[21] Held C, Cameretti LF, Sadowski G. Measuring and modeling activity coefficients in aqueous amino-acid solutions. Industrial & Engineering Chemistry Research. 2011;**50**(1):131-141

[22] Held C, Sadowski G. Thermodynamics of bioreactions. Annual Review of Chemical and Biomolecular Engineering. 2016;**7**(1):395-414

[23] Wangler A et al. Prediction and experimental validation of co-solvent influence on Michaelis constants: A thermodynamic activity-based approach. Chemistry – A European Journal. 2018. https://doi.org/10.1002/chem.201803573

[24] Prudic A, Ji Y, Sadowski G. Thermodynamic phase behavior of API/polymer solid dispersions. Molecular Pharmaceutics. 2014;**11**(7):2294-2304

[25] Reschke T, Brandenbusch C, Sadowski G. Modeling aqueous two-phase systems: I. Polyethylene glycol and inorganic salts as ATPS former. Fluid Phase Equilibria. 2014;**368** (Supplement C):91-103

[26] Gross J, Sadowski G. Application of perturbation theory to a hard-chain reference fluid: An equation of state for square-well chains. Fluid Phase Equilibria. 2000;**168**(2):183-199

[27] Rosin C et al. Combined effects of temperature, pressure, and co-solvents on the polymerization kinetics of actin. ChemPhysChem. 2015;**16**(7):1379-1385

[28] Bugnon P et al. High-pressure stopped-flow spectrometer for kinetic studies of fast reactions by absorbance and fluorescence detection. Analytical Chemistry. 1996;**68**(17):3045-3049

[29] Fuchs D et al. Solubility of amino acids: Influence of the pH value and the addition of alcoholic cosolvents on aqueous solubility. Industrial & Engineering Chemistry Research. 2006;**45**(19):6578-6584

[30] Held C, Sadowski G. Compatible solutes: Thermodynamic properties relevant for effective protection against osmotic stress. Fluid Phase Equilibria. 2016;**407**:224-235

[31] Kuby SA. A Study of Enzymes. Vol. 2. Boca Raton: CRC Press; 1990

[32] Gallati H. Stabilisierung des reduzierten β-nicotinamid-adenin-dinucleotid in einem organischen Lösungsmittel. Clinical Chemistry and Laboratory Medicine. 1976;**14**(1–12):9-14

[33] Wangler A et al. Predicting the high concentration co-solvent influence on the reaction equilibria of the ADH-catalyzed reduction of acetophenone. Journal of Chemical Thermodynamics. 2019;**128**:275-282

[34] Held C et al. ePC-SAFT revised. Chemical Engineering Research and Design. 2014;**92**(12):2884-2897

Section 2

Enzymatic Synthesis

Chapter 3

Solvent-Free Isoamyl Acetate Production via Enzymatic Esterification

Nurhazwani Yusoff Azudin and
Syamsul Rizal Abd Shukor

Additional information is available at the end of the chapter

http://dx.doi.org/10.5772/intechopen.81333

Abstract

Isoamyl acetate is an organic compound which is mainly used as flavor additive in food industries. Traditionally, the food flavor has been produced by extraction from plants, followed by chemical synthesis route which then shifted to biocatalytic route due to consumer's awareness and inclination toward natural products. This study was carried out to examine the reaction synthesis between acetic anhydride and isoamyl alcohol in the presence of *Candida antarctica* Lipase-B (CALB) as a catalyst in solvent-free system (SFS). Results show that two reactions took place between acetic anhydride and isoamyl alcohol. The effect of different reaction parameters on the final yield of isoamyl acetate and the optimization of process parameters using a statistical tool were also investigated with response surface methodology (RSM). It was found that the optimum isoamyl acetate yield is at reaction temperature 30°C, acid/alcohol molar ratio 0.10, and enzyme loading 4.14%. The regression coefficient for optimization based on RSM was 0.9961. Errors resulted from model validation is less than 1% and is acceptable for real-life application. RSM model and first principle model were selected to determine the reaction kinetics and yield of reaction for isoamyl acetate. The results showed that RSM model provides a good predication of the esterification system with R^2 value of 0.90.

Keywords: enzymatic, esterification, isoamyl acetate, solvent-free system, lipase

1. Introduction

Esters are one of the most common of all naturally occurring organic compounds which contain –COOR as functional group. Many simple esters are pleasant-smelling liquids and mainly used as fragrant odors of fruits and flowers. For example, methyl butanoate is an element found in pineapple oil, whereas isoamyl acetate is an element of banana oil [1]. These esters are also naturally present in animal fats and oil [2] and in many biologically important molecules. Esters are ubiquitous and contain "nature-identical" substance that can be used to substitute natural flavor and fragrances. The demand for flavor and fragrance products is fairly high for most applications in developed countries. In 2009, flavor and fragrance industry faced a decline due to global economic crisis, but rapidly recovered a year after. The market was forecasted to continue expanding at a CAGR of 5.6% during 2011–2013 [3].

As the demand on flavored food increased tremendously throughout the years, consumers were also concerned about the natural ingredients of it by considering food with natural flavored in their list. The term "natural" has been clearly defined by the U.S. Code of Federal Regulations 101.22(a)(3) as "…the essential oil, oleoresin, essence or extractive, protein hydrolysate (product of hydrolysis), distillate of any product of roasting, heating or enzymolysis, which contains the flavoring constituents derived from a spice, fruit juice, vegetable or vegetable juice, edible yeast, herb, bud, bark, root, leaf or similar plant material, meat, seafood, poultry, eggs, dairy products or fermentation products thereof, whose significant function in food is imparting flavoring rather than nutritional" [4].

Esters naturally available in plants and flowers were extracted for traditional flavor production. However, the traditional extraction of flavor from plants is too expensive for commercial exploitation, limitation of raw materials, and only small amount of esters produced. On the other hand, the demand of esters kept increasing; therefore, researchers overcome the problems with alternative production route via chemical synthesis. Esterification via chemical synthesis is based on Fischer esterification method. Its drawbacks attributed to the chemicals used and consumers' awareness toward chemicals added to their food makes the synthesis is not favored in the food industry. Hence, a new method of ester synthesis is required to produce large number of esters for industrial application with high economic benefit and a purer end product.

Synthesis of isoamyl acetate in organic solvent has been introduced. Due to region- and stereospecificity expressed by most lipases in mild operation conditions and high degree of purity produced, lipase-catalyzed esterification in organic solvent has recently received greater consideration relative to the traditional chemical synthetic methods, particularly in the production of natural flavor and fragrance. Despite the higher conversion yields of esters, organic solvents undoubtedly bring about negative impact on solvent toxicity, inflammable, and need extra action on separation process. In addition, some organic solvents used are too expensive to allow profitable commercial scale-up [5]. Hence, a solvent-free system was introduced in the esterification process.

The absence of solvents in the solvent-free synthesis gives advantages on the downstream processing as there would be fewer components present in the reaction mixture at the end of the esterification process. Moreover, the production cost can be minimized. In addition, Yahya et al. [6] has stated that it is possible to use high substrates' concentrations in a solvent-free system. Hence, it is scientifically and environmentally wise to produce ester via solvent-free biotechnological route that would eliminate all the disadvantages of traditional and chemical synthesis route of producing esters.

1.1. Enzymes

Enzymes react by accelerating both the rate and specificity of metabolic reactions of the substrates without changing its original shape and amount [5]. In the food industry, flavor production from enzyme-catalyzed process has been reviewed by Christen and Munguia [7]. Common types of enzymes used in food flavor are hydrolases and oxidoreductases [7, 8]. One of the hydrolases enzymes is lipase. Lipase is also capable to form ester bounds under reverse hydrolytic conditions which allow catalysis of various other types of esters [9]. The demand for natural and environmental friendly product is the main key point of the usage of lipase in ester synthesis. Therefore, lipases are considered as enzymes of high commercial potential due to their flexibility in application. There are several lipases which have been used in isoamyl acetate production. Summary on the isoamyl acetate production based on different enzymes is shown in **Table 1**.

From **Table 1**, it can be observed that a lot of works have been done by using *Candida antarctica* as compared to other lipases. Based on **Table 1**, *T. lanuginosus* has the minimum amount of ester conversion.

Recyclability of enzymes is also an important factor to be considered for industrial-scale applications as it can reduce the cost of raw materials. For esterification in the organic solvent, *S. simulans* can be reused up to 4 cycles [20] and *Rhizomucor miehei* up to 10 cycles [16], while *Candida antarctica* can be recycled more than 10 cycles [10, 14]. Previous studies also had compared the performance of enzymes toward the esterification reaction. Guvenc et al. [11] have done a study on solvent-free system and have found that Novozym 435 from *Candida antarctica* was more efficient than Lipozyme from *Rhizomucor miehei*. This is agreed by Romero et al. [13] who extended the study on kinetics of enzymatic esterification of isoamyl acetate using Novozym 435 in an organic solvent. The advantages and the benefits of using *Candida*

Lipase origins	Maximum conversion, %	References
Candida antarctica	95.5	Gubicza et al. [10]
	80	Guvenc et al. [11]
	100	Romero et al. [12]
	96	Romero et al. [13]
	100	Feher et al. [14]
	~99	Wolfson et al. [15]
Rhizomucor miehei	>90	Krishna et al. [16]
	~40	Romero et al. [13]
Mucor miehei	>80	Krishna et al. [17]
	96.4	Mittelbach and Trathnigg [18]
S. simulans	64	Ghamgui et al. [19]
T. lanuginosus	~42	Romero et al. [13]

Table 1. Enzymes and its corresponding maximum conversion of substrates based on previous studies.

antarctica in isoamyl acetate production in solvent-free system, as listed in previous research works, provide a sound basis of choosing this lipase in this present study.

2. Materials and methods

2.1. Materials and chemicals

In this study, isoamyl acetate was produced experimentally by reacting acetic anhydride and isoamyl alcohol with the presence of enzyme, *Candida antarctica* Lipase-B (CALB) in a solvent-free system. The chemicals used in this study were analytical grades and are summarized in **Table 2** together with the respected purity, usage, and supplier. The chemicals were used as received without further purification.

2.2. Equipment

The production of isoamyl acetate enzymatic synthesis from acetic anhydride was done in lab scale. All of the experimental works were carried out using 100-ml Erlenmeyer flasks with stopped rubber, which then were placed in an incubator shaker (Benchmark Incu-shaker mini, New Jersey). Incubator shaker was used to maintain the mixing rate and to control the temperature. Then, flame ionization detector gas chromatography (GC-FID) (Agilent Technologies, 7820A GC system, USA) was used to analyze the concentrations of compounds in the sample taken.

2.3. Isoamyl acetate syntheses

Isoamyl acetate syntheses were carried out without any organic solvent in 100-ml stopped-rubber Erlenmeyer flask with working volume of 15 ml. Enzyme was added into the reaction media containing a mixture of isoamyl alcohol and acetic anhydride at various temperatures. The reaction mixture was then incubated in an incubator shaker (Benchmark) at 150 rpm for 6 h. The basis of this experimental method was taken from [16].

2.4. Analysis of esterification

About 0.5 ml of the reaction mixture was withdrawn periodically starting from t = 0 h, until t = 6 h for analysis. The withdrawal was done using micropipette and transferred into

Materials/chemicals	Purity	Usage	Supplier
Isoamyl alcohol	99.8%	Production medium	Merck Co., Malaysia
Acetic anhydride	98%	Production medium	ACROS Organics, Malaysia
Isoamyl acetate	100%	GC standard	Merck Co., Malaysia
CALB (≥5000 U/g)	N/A	Production medium	Sigma-Aldrich, Malaysia

Table 2. List of materials and chemicals used.

microcentrifuge tube. Samples were analyzed using gas chromatograph (Agilent Technologies 7820A) equipped with a hydrogen flame ionization detector and a SGE BP21 (FFAP) column (60 m × 0.32 mm × 0.25 μm). Helium was used as a carrier gas at a flow rate of 5 ml/min. After injection of samples, the oven temperature was kept at 100°C and linearly increased to 140°C. The rate of temperature increase was set at 70°C/min, and was kept at 140°C for the remaining time of analysis. Injector and detector temperatures were set at 200 and 250°C, respectively.

Quantification of data was done by calibration with standards samples. Each sample required 4.08 min to be analyzed by GC-FID. The retention times of peaks were as follows: isoamyl acetate, 2.26 min; isoamyl alcohol, 2.38 min; acetic anhydride, 2.48 min; and acetic acid, 3.2 min.

2.5. Effect of reaction temperature

The effects of reaction temperature on the enzymatic esterification were studied at various temperatures: 30, 40, and 50°C. About 15 ml working volume of the medium in a 100-ml Erlenmeyer flask was incubated in an incubator shaker with agitation speed of 150 rpm for 6 h reaction time. Samples were taken periodically until 6 h of reaction time and analyzed by GC-FID for isoamyl acetate production.

2.6. Effect of acid/alcohol molar ratio

The effect of acid/alcohol molar ratio was studied at various acid/alcohol molar ratios: 0.1 (excess alcohol), 1 (equimolar), and 2 (excess acid). The medium was incubated in an incubator shaker at 40°C reaction temperature, and with agitation speed of 150 rpm for 6 h reaction time. Samples were taken periodically until 6 h of reaction time and analyzed with GC-FID for isoamyl acetate content.

2.7. Effect of enzyme loading

The enzyme loading effects were studied at various percentages of enzymes in medium: 4, 8, and 12%. The calculation was based on the overall mass of substrates used in the reaction. The medium together with the enzyme was incubated in an incubator shaker at a temperature of 40°C. The agitation speed was set to 150 rpm; samples were taken at different time intervals until 6 h of reaction time and were analyzed using GC-FID for isoamyl acetate content.

2.8. Optimization process using response surface methodology (RSM)

Optimization studies are carried out using response surface methodology (RSM). RSM is a collection of statistical and mathematical analysis for developing, improving, and optimizing processes in which the response developed is influenced by several variables. It has an important application in the process development, design, and formulation of new products, as well as in the improvement of existing product design.

			Levels		
Variables	Coding	Unit	−1	0	+1
Temperature	A	°C	30.00	40.00	50.00
Ac/Al ratio	B	—	0.1	1	2
Enzyme loading	C	%	4	8	12
Reaction time	D	h	2	4	6

Table 3. List of variables and its value.

2.9. Model fitting and statistical analysis

Optimum conditions for isoamyl acetate enzymatic esterification can be obtained by using optimization software, Design Expert 6.0.6. The method used was Central Composite Design (CCD), under RSM. CCD is the best design for response optimization [21]. In this study, three levels and four factor variables were chosen. The three levels represent the three points between the lower and upper limit of the parameters, whereas the four factors represent the four parameters that are studied in this section, which are reaction temperature, ac/al molar ratio, enzyme loading, and reaction time. The details of the parameters and levels studied were shown in **Table 3**.

2.10. Sensitivity analysis

Sensitivity analysis is useful for testing the robustness of the result of a model developed, to show the relationships between input and output variables in a system, and for model simplification by removing the insensitive or insignificant variables.

Based on the optimization step before, a sensitivity analysis for each individual parameter and the interaction between parameters in this study were done using application tools provided by RSM. It was done to decide the interaction between parameters, and the most sensitive parameters in esterification process.

3. Enzyme kinetic modeling

3.1. Response surface methodology (RSM) model

RSM model was developed using DoE software by designing new experiments for enzyme kinetic model. Three-level and four-factor designs which consist of 27 sets of experiment were constructed using parameters listed in **Table 4**.

3.2. First principle model

First principle model is an application of conservation of mass to the analysis of a physical system by taking account of material entering, leaving, generating, consuming, and accumulating in the system.

Variables	Coding	Unit	Levels		
			−1	0	+1
Temperature	β_1	°C	30	40	50
Mass enzyme	β_2	wt%	4	8	12
Reaction time	β_3	h	2	4	6
Reciprocal of anhydride concentration	β_4	l/mol	0.12	0.20	1.18

Table 4. Experimental range and levels of variables.

The mathematical general equation of a balanced mass conservation quantity by using conservation law in a system is:

$$Input + Generation = Output + Accumulation \tag{1}$$

In term of general mole balance, the above equation became

$$Q_0 C_{j0} + \int_V r_j dV = \frac{d}{dt}\int_V C_j dV + Q_1 C_{j1} \tag{2}$$

By assuming component j enters and leaves the element only by convection with the inflow and outflow streams by neglecting diffusional flux through the boundary of the volume element, Q_0 and Q_1 are the mass of component j at the inflow and outflow, respectively, C_{j0} and C_{j1} are the concentration of component j at the inflow and outflow, respectively, $\int_V r_j dV$ is the rate of generation of component j, and $\frac{d}{dt}\int_V C_j dV$ is the rate of accumulation of component j.

In a close system (batch process), assuming with well stirred substrate, the above equation reduces to:

$$Generation = Accumulation \tag{3}$$

Therefore,

$$\int_V r_j dV = \frac{d}{dt}\int_V C_j dV \tag{4}$$

where r is the rate of reaction, V is the volume, and C_j is the concentration of product produced by time in the reaction system. Since the volume of the reactor is constant in batch system, thus Eq. (4) reduces to:

$$r_j = \frac{dC_j}{dt} \tag{5}$$

The reaction rates can be derived in detail by enzyme kinetic equation.

3.3. Validation of kinetic model

The entire model developed will then need to be validated to assure the models are reliable and can be used in industrial application. Validation of the kinetic model was done by comparing the output from the model developed with the experimental data collected. The results were compared and plotted in a graph, and error value based on the regression analysis was done.

4. Results and discussions

4.1. Isoamyl acetate syntheses

Theoretically, acetic anhydride possesses two acyl groups. In a reaction with isoamyl alcohol, one of the acyl from acetic anhydride will bind with isoamyl alcohol and discharge one H^+ to form isoamyl acetate and acetic acid. Then, excess acyl (from acetic acid) will react with excess alcohol to form another isoamyl acetate and water. The details of reaction scheme are shown below:

a. First reaction

$$Isoamyl\ alcohol + Acetic\ anhydride \rightarrow Isoamyl\ acetate + Acetic\ acid$$

b. Second reaction

$$isoamyl\ alcohol + acetic\ acid \rightarrow isoamyl\ acetate + water$$

c. Overall reaction

$$isoamyl\ alcohol + acetic\ anhydride \rightarrow isoamyl\ alcohol + water$$

Based on the chemical reaction scheme shown previously, esterification of isoamyl acetate undergo two reactions, first is between the acetic anhydride and isoamyl alcohol, producing acetic acid and isoamyl acetate; and the second reaction is between the acetic acid and excess isoamyl alcohol, producing another isoamyl acetate and water as by-product. An overall analysis on the effect of substrates concentration was done at 8% enzyme loading, 6 h of reaction time at acid/alcohol ratio of 0.1, 1, and 2, and the results were plotted in **Figures 1–3** respectively.

From **Figure 1**, the concentration of isoamyl acetate increased rapidly at the initial of the reaction until 15 min of reaction time. During that time, acetic anhydride was consumed until 98% of its initial concentration, whereas the concentration of acetic acid and isoamyl acetate was increasing. This is in line with the reaction mechanism involved, where the reaction between acetic anhydride and isoamyl alcohol will produce acetic acid and isoamyl acetate initially. As clearly shown in **Figure 1**, as acetic anhydride was completely consumed, acetic acid produced will then react with the excess isoamyl alcohol, producing isoamyl acetate and water. This is evident by the reducing of acetic acid concentration after 15 min of reaction time and consequently the rapid increase of isoamyl acetate concentration. Throughout that time,

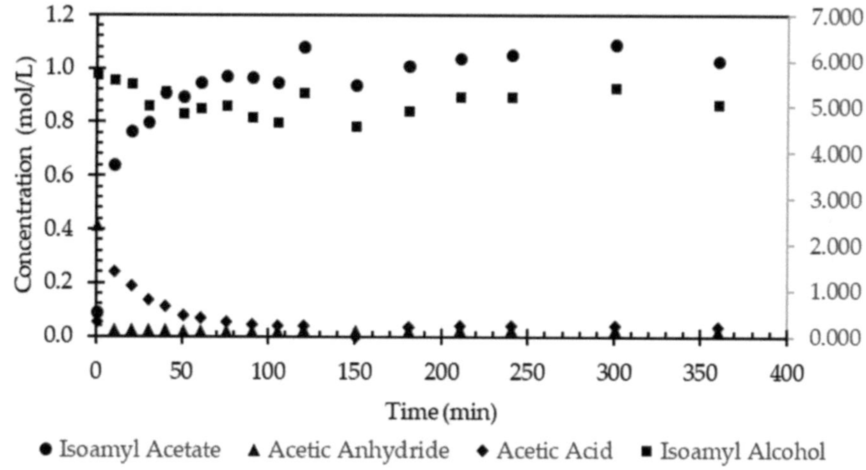

Figure 1. Overall concentration of substrates and product during the reaction for acid/alcohol ratio of 0.1.

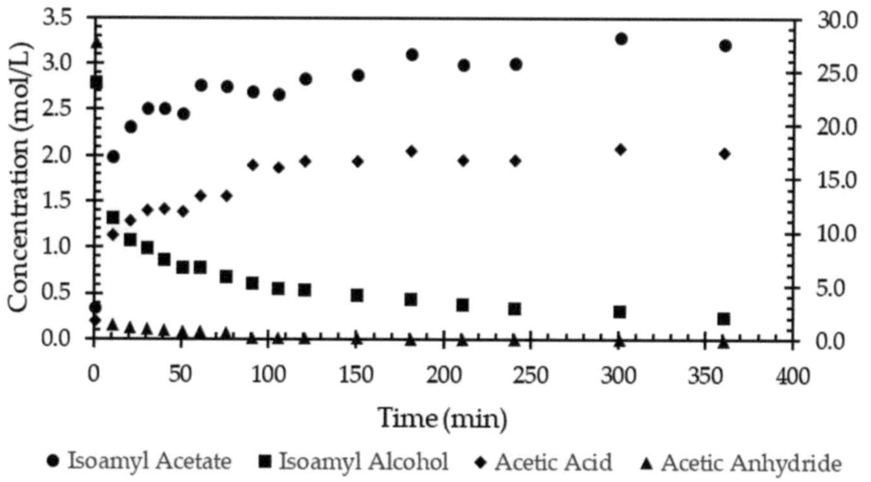

Figure 2. Overall concentration of substrates and product during the reaction for acid/alcohol ratio of 1.0.

concentration of acetic acid starts decreasing until 80% of acetic acid's initial production. At the end of the 6-h reaction time, isoamyl alcohol was in excess, acetic anhydride and acetic acid were 98 and 80% consumed, respectively.

From **Figure 2** initially, concentration of isoamyl acetate increased rapidly at the first 30 min of reaction time. During that time, acetic anhydride was consumed until 95% of its initial concentration, whereas concentration of acetic acid and isoamyl acetate was increasing, following the reaction mechanism described by the above reaction equation, which is reaction between acetic anhydride and isoamyl alcohol will initially produce acetic acid and isoamyl acetate. Acetic anhydride was fully consumed during 180 min of reaction time, and acetic acid produced at the beginning of reaction was in excess until the end of 6 h reaction duration as well as isoamyl

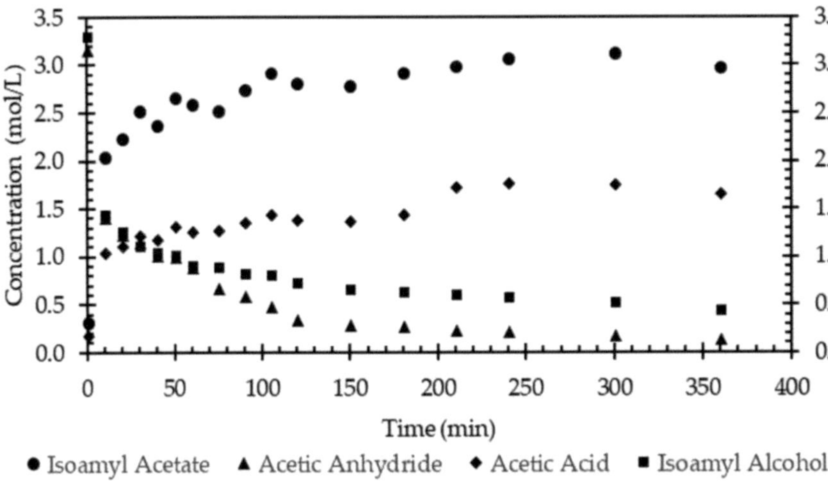

Figure 3. Overall concentration of substrates and product during the reaction for acid/alcohol ratio of 2.0.

alcohol. This is because the production of isoamyl acetate has achieved steady state, and the excess amount of alcohol is lower than the minimum amount which required to be reacted.

Figure 3 clearly shows that in excess acetic anhydride condition, isoamyl alcohol was 98% consumed during the 240 min of reaction time. After that, there is no decreasing trend in either acetic anhydride or acetic acid concentration. Acetic acid produced by the primary reaction at the beginning of the test was in excess since isoamyl alcohol is in limited condition. This shows that both reactions were stopped by limited amount of isoamyl alcohol in the mixture.

Based on **Figures 1–3**, isoamyl alcohol is the critical substrate in this esterification reaction. The final amount of isoamyl acetate produced is highly depending on the amount of isoamyl alcohol at the initial of the reaction. Limited amount of isoamyl alcohol will automatically stop the reaction mechanism, since there is no receiver of acyl in the mixtures. This also shows that acetic anhydride amount in the mixtures need to be restricted to some amount so that there is no excess acyl in the reaction, and hence making this process cost-effective [22].

4.2. Effect of synthesis parameters

4.2.1. Effect of reaction temperature

Reaction temperature of esterification process has an effect on the final yield of isoamyl acetate produced based on the Arrhenius equation (Eq. (6)).

$$k = k_0 \, exp\left\{-\frac{E_a}{RT}\right\} \tag{6}$$

where k is the rate constant, k_0 is the pre-exponential constant, E_a is the activation energy, R is the gas constant, and T is the absolute temperature.

Eq. (6) clearly shows that reaction temperature has parallel effect to the reaction rate constant and hence influences the final yield of isoamyl acetate produced by affecting the esterification reaction rate.

Effect of temperature on the yield of ester produced has been investigated at temperatures 30, 40, and 50°C for 6 h of reaction time. From literatures, the range of temperature studied was between 30 and 65°C. However, the optimum reaction temperature was found to be between 30 and 50°C. Hence, to elucidate the impact of reaction temperature on yield of ester, the synthesis has been studied at a temperature range from 30 to 50°C for 6 h of reaction time, at 8% enzyme concentration and acid/alcohol molar ratio of 1. The results are illustrated in **Figure 4**.

Based on **Figure 4**, initial reaction rate of ester production increased with increasing reaction temperature from 30 to 50°C. This would be explained by Eq. (6), where increasing reaction temperature would increase the kinetics of the reaction, hence encourage the collision rate between molecules in the medium, and thus favor higher production of ester. This result is in agreement with a research done by [23, 24], where increasing reaction temperature will increase the reaction rate and hence produce higher concentration of ester. Similar result on the positive effect of reaction kinetics toward the increasing of reaction temperature was also found by the studies.

As the reaction time increased, the production rate of ester appears to decrease for reaction temperatures of 40 and 50°C compared to the production at 30°C. The final yield of ester produced at 30, 40, and 50°C of reaction temperatures at time 6 h of reaction time were 67.2, 61.8, and 59.1%, respectively. This could be due to the enzyme tertiary structure that starts to disrupt at higher reaction temperature and at longer reaction time, hence losing its catalytic activity, thus lowering the enzyme production rate [25].

4.2.2. Effect of acid/alcohol molar ratio

The effect of acid/alcohol molar ratio has been studied at low anhydride concentration, equimolar, and excess in anhydride, whose ratios were 0.1, 1, and 2, respectively. The reaction

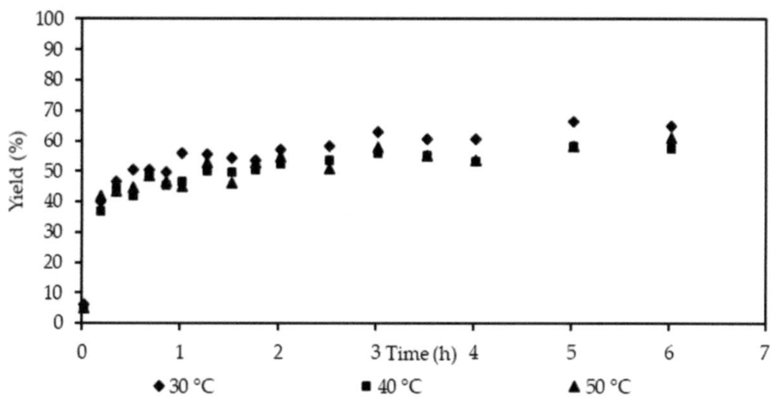

Figure 4. Effect of temperature on the ester production.

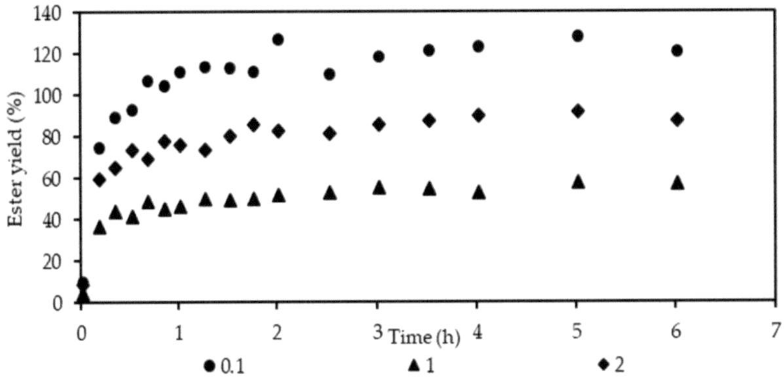

Figure 5. Effect of acid-alcohol ratio on the yield of ester.

temperature was set at 40°C and enzyme loading at 8%. The result for this experimental condition is shown in **Figure 5**.

Based on **Figure 5**, a maximum yield of isoamyl acetate was achieved when isoamyl alcohol is in excess over the acetic anhydride ratio. This is due to the availability of excess nucleophile in the reaction mixture. Based on the reaction scheme shown in Section 4.1, an acyl from acetic anhydride is reacted with a nucleophile available from isoamyl alcohol and producing isoamyl acetate and acetic acid. Then, an acyl from acetic acid produced reacted for the second time with the available nucleophile in the reaction mixture and produced another isoamyl acetate and water. The maximum yield of isoamyl acetate produced at acid/alcohol molar ratios of 0.1, 1, and 2 was 128, 53, and 88%, respectively.

The current study is in agreement with [14] and [26], where higher substrate concentration (high acid/alcohol molar ratio) leads to lower yields of ester. This could also be due to acid concentration which has met the critical concentration needed in the reaction [27]. However, by increasing the molar ratio to excess acid, the yield of ester starts to increase again. This condition happened because the concentration of acid in the reaction mixture has past the critical concentration of acid needed in the reaction medium.

4.2.3. *Effect of enzyme loading*

The effect of enzyme loading concentration was studied at 4, 8, and 12% of enzyme loading, acid/alcohol molar ratio of 1 and 30°C of reaction temperature. Enzyme loading is economically important to the esterification process as enzyme is costly compared to other materials used in the synthesis. Producing high yield of isoamyl acetate at low quantity of enzyme synthesis was highly preferred in an esterification synthesis. Therefore, in this study, the yield of isoamyl acetate produced versus reaction time was plotted and shown in **Figure 6** for three different enzyme loadings.

Based on the figure above, increasing amount of enzyme loading increases the yield of ester produced. The maximum yield of isoamyl acetate produced by the synthesis for 4, 8, and 12%

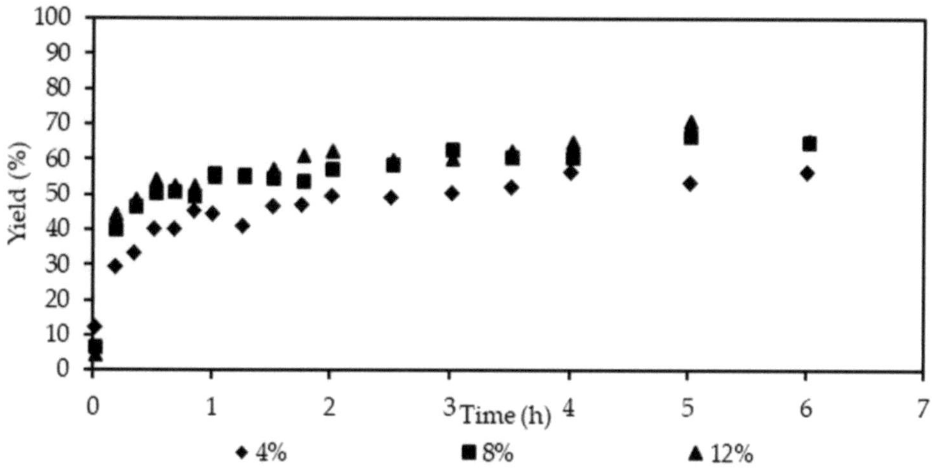

Figure 6. Effect of enzyme loading on the ester production.

are 57, 65.8, and 71.5%, respectively. The graph also shows the clearer trend of enzyme loading effect along the 6 h of reaction time. For lower amount of enzyme loading, longer reaction needed to achieve steady state of the esterification reaction. Increasing the enzyme loading from 4 to 12% shortened the time required for the reaction to achieve steady state which is favorable. It also shows that esterification reaction increases with the increasing amount of enzyme loading and the trends resembles the results of esterification studied by [26]. This concludes that the rate of esterification is dependent on the enzyme concentration used in the mixture.

4.3. Optimization using response surface methodology (RSM)

Design Expert software contained many types of optimization methods that can be used in this study. Based on the optimization methods described, a central composite design (CCD) method was chosen to design all the experiments conducted for esterification synthesis. CCD is generally the best design for response optimization, as stated by [21, 28]. In this study, three-level and four-factor designs were used to determine the optimum condition with four experimental parameters (temperature, acid-alcohol molar ratio, enzyme ratio, and reaction time) and three-level indicated the level of each range (−1, 0, +1). Six replicates which run at the center point (0, 0) of the design were performed to allow the estimation of pure error. All the experiments were carried out in the randomized order to minimize the unexplained variability in the observed responses due to irrelevant factor.

Fitting of the data to models in RSM and their subsequent ANOVA showed by **Figure 7** confirmed that the enzymatic esterification reaction of acetic anhydride and isoamyl alcohol was most suitably described by quadratic model. The equation of the model based on the actual values is shown by Eq. (7).

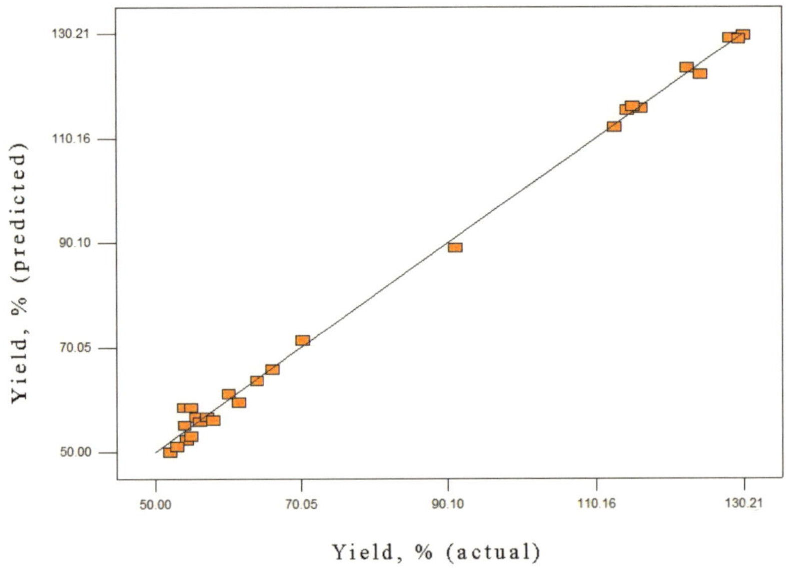

Figure 7. Parity plot for relation between observed and predicted isoamyl acetate synthesis.

$$\begin{aligned} yield = {} & 271.8778 - 11.0681\boldsymbol{A} - 303.3621\boldsymbol{B} + 9.9557\boldsymbol{C} \\ & - 11.5634\boldsymbol{D} + 0.1459\boldsymbol{A}^2 + 36.25\boldsymbol{B}^2 - 0.4381\boldsymbol{C}^2 - 1.2452\boldsymbol{D}^2 \\ & + 10.2656\boldsymbol{AB} - 0.0588\boldsymbol{AC} + 0.3626\boldsymbol{AD} + 0.4859\boldsymbol{BC} \\ & - 8.678\boldsymbol{BD} - 0.2537\boldsymbol{CD} - 0.1443\boldsymbol{A}^2\boldsymbol{B} + 0.0011\boldsymbol{A}^2\boldsymbol{C} \\ & - 0.0072\boldsymbol{A}^2\boldsymbol{D} + 0.3997\boldsymbol{AB}^2 - 0.0222\boldsymbol{ABC} + 0.2123\boldsymbol{ABD} \\ & - 0.0037\boldsymbol{ACD} + 0.0151\boldsymbol{BCD} \end{aligned} \tag{7}$$

The quadratic response function represents the yield of ester produced, where **A** is reaction temperature, **B** is the acid/alcohol molar ratio, **C** is percentage of enzyme used, and **D** is the reaction time.

Closer the value of R^2 to unity, the better the empirical models fit the experimental data. Parity plot between actual data and the predicted data is done and shown in **Figure 7**. As can be seen, the predicted values match the actual values reasonably well within the ranges of the experimental parameter conditions, with R^2 value of 0.9961. At this stage, this result suggests the applicability and reliability of the equation in representing the esterification reaction between acetic anhydride and isoamyl alcohol by using CALB in a solvent-free system with sufficient degree of accuracy.

Statistical analysis from the analysis of variance (ANOVA) was done using RSM software (result not shown). The *F* value of the model (47.022) with a *P* value of 0.001 implied that the model is significant. Generally, *P* value lower than 0.01 indicates that the model is considered to be statistically significant at 99% confidence level and values greater than 0.10 indicate that the terms are not significant [29]. Based on the results, the most significant model terms (*F* value = 23.014) that give high impact on the yield of ester produced is the acid/alcohol molar ratio, with *P* value of 0.0087. The small *P* value (<0.001) and a high regression coefficient ($R^2 = 0.9961$) showed the suitability of the model for representing the real relationship between all the reaction parameters and yield of the ester produced. Adequate precision value measured the signal-to-noise ratio for the model. Ratios greater than 4 indicated adequate model discrimination [28]. In this study, the adequate precision for developed model was found to be 17.6666; this indicates that the model could be used to navigate the design space for this enzymatic esterification reaction.

Model developed by RSM needs to be validated to make sure it is reliable and acceptable. **Figure 8** shows the comparison of ester yield from the actual experiment and calculated by model developed. The comparison was done for 27 runs of experiments.

Based on **Figure 8**, the yield of ester's model developed by RSM appear to fit well with the actual results of ester yield from the experimental data. This clearly shows that the model developed is reliable to use as prediction of the real process of enzymatic esterification synthesis between acetic anhydride and isoamyl alcohol catalyzed by CALB in solvent-free system.

4.3.1. Numerical optimization

Numerical optimization was done by setting the desired optimum condition and product, and then DoE software will generate few solutions based on the model developed before. Based on the desired optimized condition setting, four sets of solutions for optimized experimental condition for enzymatic esterification in SFS were developed as shown in **Table 5**.

Based on the numerical optimization, the most desirable reaction conditions for optimum isoamyl acetate yield was set at minimum reaction temperature, enzyme loading, and reaction time. Minimum reaction temperature is required to reduce the energy usage, hence lower the

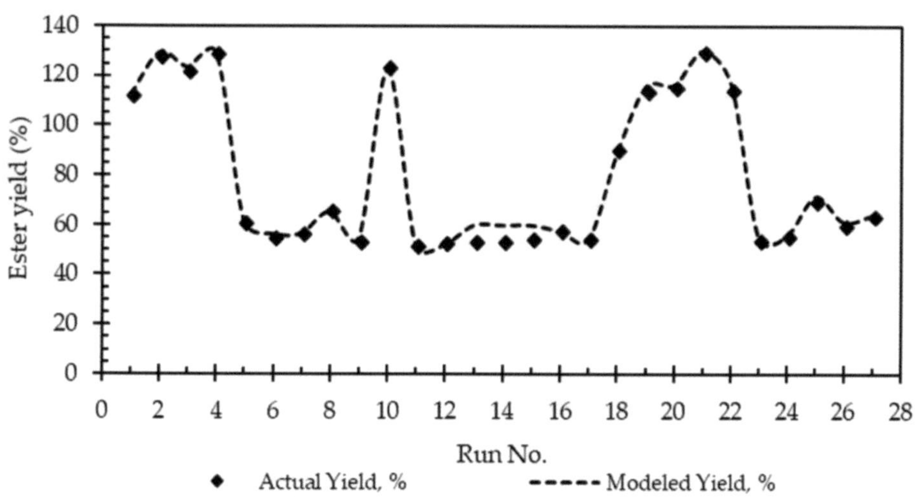

Figure 8. Comparison between the RSM model developed and the actual experimental data.

Parameters	Constrains			Solution
	Goal	Lower limit	Upper limit	
T (°C)	Minimize	30	50	30
Ac/Al	Is in range	0.1	2	0.10
Enzyme (%)	Minimize	4	12	4.14
Time (h)	Minimize	2	6	2.00
Yield (%)	Maximize	52.00	130.21	112.83

Table 5. Optimized condition suggested by DoE software.

operating cost of esterification process. Enzyme loading was set to minimum in order to reduce the usage of enzyme in esterification synthesis as enzyme is the most expensive material in esterification synthesis. Lowering reaction time to the minimum can help to accelerate the production rate of isoamyl acetate and give benefit to the industry. The results for optimum condition are shown in **Table 6**.

Experimental result showed that there was no significant difference for percentage yield of ester produces between the predicted and actual values. The error value for the optimization experiment is less than 1%. Therefore, the model obtained is reliable to predict the yield of isoamyl acetate in enzymatic esterification of isoamyl acetate by solvent-free system with high accuracy.

4.4. Enzyme kinetic

4.4.1. Response surface methodology (RSM) model

Design of experiment (DoE) software is used again to design the experiment. Three-level and four-factor designs which consist of 27 sets of experiments were done. The effect of temperature, *T*, mass of enzyme, *M*, reaction time, *t*, and reciprocal of initial anhydride concentration, $1/[A]_0$, on the enzymatic reaction rate were investigated using CCD analysis in RSM.

All coefficients obtained from the full quadratic polynomial model were evaluated by regression analysis and tested for their significance. The insignificant coefficients were eliminated based on *p* values. It was found that coefficients for β_4^2 and β_{23} were highly insignificant, hence the predicted polynomial model was rearranged by eliminating the terms which consist of β_4^2 and β_{23}, resulting in a modified quadratic model. The coefficient of determination ($R^2 = 0.99$) implies that the model was satisfactory.

The final model for reciprocal of enzymatic reaction rate obtained from the CCD analysis is:

Number	Temperature (°C)	Ac/Al ratio	Enzyme loading (%)	Reaction time (h)	Yield (%)		Error (%)
					Predicted	Actual	
1	30.0	0.10	4.14	2.0	112.83	112.63	0.18

Table 6. Result and error analysis of model validation.

$$\frac{1}{r} = 7.69399 - 0.2021\,(T) - 0.3193(M) - 1.0334(t) + 10.0771\left(\frac{1}{[A]_0}\right) + 0.028\left(T^2\right) + 0.0239\left(M^2\right) + 0.1060\left(t^2\right) - 0.024(TM) + 0.0385\,(Tt) - 0.1342\left(T\left(\frac{1}{[A]_0}\right)\right) - 0.0732\left(M\left(\frac{1}{[A]_0}\right)\right) - 0.1623\left(t\left(\frac{1}{[A]_0}\right)\right) \quad (8)$$

4.4.2. First principle model

By using Eq. (8), the general equation for reaction rate developed throughout the experiment is given by:

$$\frac{dC_j}{dt} = r_j \quad (9)$$

Given that general reaction rate equation, r is;

$$r_j = kC_j^n \quad (10)$$

where r_j is the reaction rate for product j, k is the reaction rate constant, C_j is the concentration of product j, and n is the order of reaction.

Since the overall reaction is reversible, hence Eq. (10) becomes

$$r_j = k_1 C_j^n - k_2 C_i^n \quad (11)$$

where k_1 and k_2 are rate constants for forward and backward reaction, respectively, C_i and C_j are concentration for substrates i and product j, respectively, and n is the order number of the reaction.

Overall reaction for isoamyl acetate esterification from acetic anhydride and isoamyl alcohol is:

$$2CH_3CH(CH_3)CH_2CH_2OH + (CH_3CO)_2O \underset{k_2}{\overset{k_1}{\rightleftharpoons}} 2CH_3COOCH_2(CH_3)CHCH_3 + H_2O$$

Based on Eq. (11), there were few reaction rate equations that are possible to be applied for this isoamyl acetate enzymatic esterification reaction; therefore, all of the possible reaction rate equations were listed in **Table 7** and reaction rate constants for all equation developed were solved out using nonlinear equation solver in POLYMATH and the regression of each equation was compared. **Table 8** shows that the accurate reaction rate equation was for the third reaction rate with $R^2 = 0.9385$ and $adj.R^2 = 0.9360$. The kinetic constant for k_1 and k_2 equals to -0.0135 and 0.2530, respectively, with order number of 1. Therefore, the final reaction rate equation for enzymatic esterification reaction from acetic anhydride and isoamyl acetate becomes

$$r = -0.0135C_b + 0.2530C_p \quad (12)$$

where C_b is the concentration of isoamyl alcohol and C_p is the concentration of ester.

No.	Reaction rate equation	k_1	k_2	R^2	$Adj.R^2$
1	$r = (k_1C_a) + (k_2C_c)$	0.0461	0.3984	0.5705	0.5533
2	$r = (k_1C_a) + (k_2C_p)$	0.0078	0.2426	0.9121	0.9086
3	$r = (k_1C_b) + (k_2C_p)$	−0.0135	0.2530	0.9385	0.9360
4	$r = (k_1C_a^2) + (k_2C_p)$	0.0013	0.2447	0.9109	0.9073
5	$r = (k_1C_a^2) + (k_2C_p^2)$	0.0187	0.0912	0.8111	0.8035
6	$r = (k_1C_b^2) + (k_2C_p^2)$	0.0032	0.0961	0.7864	0.7778
7	$r = (k_1C_b) + (k_2C_p^2)$	0.0172	0.0951	0.7890	0.7806
8	$r = (k_1C_aC_b) + (k_2C_p^2)$	0.1912	0.0909	0.7936	0.7853
9	$r = (k_1C_aC_b) + (k_2C_p)$	−0.0147	0.2474	0.9109	0.9073

Table 7. Possible reaction rate equation, constant developed, and regression analysis.

Model	R^2
RSM model	0.90
First principle model	0.89

Table 8. Regression analysis between the model developed and the actual data from experimental results.

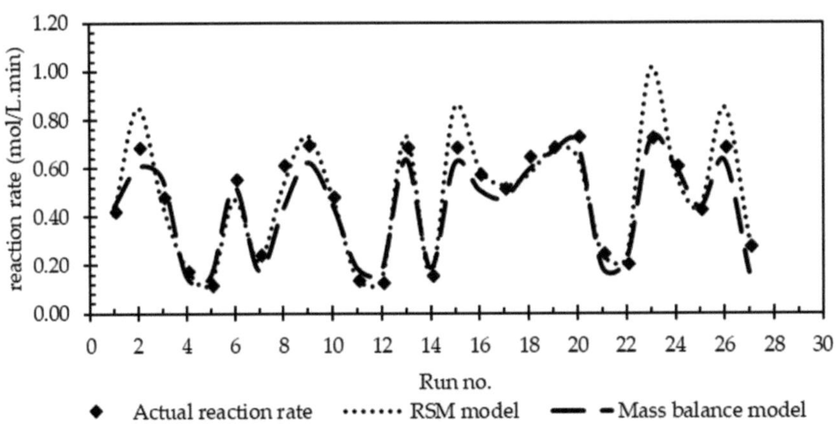

Figure 9. Comparison between the actual reaction rate and the modeled reaction rate.

4.4.3. Validation of kinetic equation

Validation of kinetic equation was done by comparing the experimental data for reaction rate of each run with the reaction rate calculated by Eqs. (8) and (12). Then, the results were plotted in **Figure 9** and regression analysis was done to evaluate the accuracy of models developed.

Regression analysis was done between the models and the actual data by using excel and the result of regression analysis is shown in **Table 8**. The regression value (R^2) calculated for RSM model is 0.90, and for first principle model is 0.89. Based on **Figure 9** and **Table 8**, first principle model and RSM model was found to have good agreement with the actual data from the experiment. Hence, kinetics of enzymatic esterification of isoamyl acetate in this study can be represented by RSM and mass balance kinetic model.

5. Conclusions

In this study, it was concluded that there are two chemical reactions involved in the esterification of isoamyl acetate from acetic anhydride and isoamyl alcohol. The main reaction is between acetic anhydride and isoamyl alcohol, and the combination of the two reactions results in an overall reaction as follow:

$$2CH_3CH(CH_3)CH_2CH_2OH + (CH_3CO)_2O \rightleftharpoons 2CH_3COOCH_2(CH_3)CHCH_3 + H_2O$$

Between all of the parameters studied, the most critical parameter in isoamyl acetate synthesis is the acid/alcohol molar ratio. This is because the molar ratio will affect the nucleophile and acyl content in the mixture. The least sensitive parameter is the reaction temperature of the ester synthesis. This can be shown by small gap on the yield of ester produced at large different of reaction temperature.

Optimization of enzymatic isoamyl acetate synthesis in solvent-free system was done using RSM. The present process fits well with second order quadratic equation with determination of coefficient (R^2) equals to 0.9961. The numerical optimization suggested that the optimum condition for enzymatic esterification of isoamyl acetate from acetic anhydride and isoamyl alcohol by enzyme CALB in solvent-free system is at 30°C reaction temperature, 0.10 acid/alcohol molar ratio, 2 h of reaction time, and 4.14% of enzyme loading. Two types of mathematical models were selected and validated to determine the reaction kinetics and yield of reaction for isoamyl acetate. From two models selected, the RSM model is the most accurate model for the esterification with R^2 value of 0.90.

Acknowledgements

Author would like to acknowledge Ministry of Education Malaysia for the financial support through MyBrain15 scheme.

Author details

Nurhazwani Yusoff Azudin and Syamsul Rizal Abd Shukor*

*Address all correspondence to: chsyamrizal@usm.my

School of Chemical Engineering, Engineering Campus, Universiti Sains Malaysia, Nibong Tebal, Penang, Malaysia

References

[1] Mc Murry J. Organic Chemistry. 7th ed. Thomson Learning, Inc: USA; 2008

[2] Shaw J-F, Wu H-Z, Shieh C-J. Optimized enzymatic synthesis of propylene glycol monolaurate by direct esterification. Food Chemistry. 2003;**81**:91-96. DOI: 10.1016/s0308-8146(02)00383-7

[3] Koncept Analytics. Global Flavor & Fragrance Market Report—2011. Edition 2011. 2012. Available from: http://www.marketresearch.com/Koncept-Analytics-v3494/Global-Flavor-Fragrance-Edition-6738031/

[4] Food Drug Administration Center. Food and Drugs: Food for general consumption. US Food and Drug Administration. 2017:21-25

[5] Grisham CM, Garrett RH. Biochemistry. Philadelphia: Sounders College Pub. hlm; 1999

[6] Yahya ARM, Anderson WA, Moo-Young M, Moo-Young ME. Ester synthesis in lipase-catalyzed reactions. Enzyme and Microbial Technology. 1998;**23**:438-450

[7] Christen P, Munguia AL. Enzymes and food flavor—A review. Food Biotechnology. 1994;**8**:167-190

[8] Kuntz LA. Enzyme that Aid Beverages. 1996. Available from: http://www.foodproductdesign.com/articles/1996/09/enzymes-that-aid-beverages.aspx

[9] Divankar S, Manohar B. Use of lipases in the industrial production of esters. Industrial Enzymes. 2007:283-300

[10] Gubicza L, Kabiri-Badr A, Keoves E, Belafi-Bako K. Large-scale enzymatic production of natural flavour esters in organic solvent with continuous water removal. Journal of Biotechnology. 2000;**84**:193-196

[11] Guvenc A, Kapucu N, Mehmetoğlu Ü. The production of isoamyl acetate using immobilized lipases in a solvent-free system. Process Biochemistry. 2002;**38**:379-386. DOI: 10.1016/s0032-9592(02)00099-7

[12] Romero MD, Calvo L, Alba C, Habulin M, Primožič M, Knez Ž. Enzymatic synthesis of isoamyl acetate with immobilized *Candida antarctica* lipase in supercritical carbon dioxide. Journal of Supercritical Fluids. 2005;**33**:77-84. DOI: 10.1016/j.supflu.2004.05.004

[13] Romero MD, Calvo L, Alba C, Daneshfar A. A kinetic study of isoamyl acetate synthesis by immobilized lipase-catalyzed acetylation in n-hexane. Journal of Biotechnology. 2007;**127**:269-277. DOI: 10.1016/j.jbiotec.2006.07.009

[14] Feher E, Illeova V, Kelemen-Horvath I, Belafi-Bako K, Polakovic M, Gubicza L. Enzymatic production of isoamyl acetate in an ionic liquid-alcohol biphasic system. Journal of Molecular Catalysis B: Enzymatic. 2008;**50**:28-32

[15] Wolfson A, Atyya A, Dlugy C, Tavor D. Glycerol triacetate as solvent and acyl donor in the production of isoamyl acetate with *Candida antarctica* lipase B. Bioprocess and Biosystems Engineering. 2010;**33**:363-366

[16] Krishna SH, Divakar S, Prapulla SG, Karanth NG, Hari Krishna S, Divakar S, et al. Enzymatic synthesis of isoamyl acetate using immobilized lipase from *Rhizomucor miehei*. Journal of Biotechnology. 2001;**87**:193-201. DOI: 10.1016/s0168-1656(00)00432-6

[17] Hari Krishna S, Manohar B, Divakar S, Prapulla SG, Karanth NG, Krishna SH, et al. Optimization of isoamyl acetate production by using immobilized lipase from *Mucor miehei* by response surface methodology. Enzyme and Microbial Technology. 2000;**26**:131-136. DOI: 10.1016/S0141-0229(99)00149-0

[18] Mittelbach M, Trathnigg B. Kinetics of alkaline catalyzed methanolysis of sunflower oil. Lipid/Fett. 1990;**92**:145-148. DOI: 10.1002/lipi.19900920405

[19] Ghamgui H, Karra-Chaâbouni M, Bezzine S, Miled N, Gargouri Y. Production of isoamyl acetate with immobilized *Staphylococcus simulans* lipase in a solvent-free system. Enzyme and Microbial Technology. 2006;**38**:788-794. DOI: 10.1016/j.enzmictec.2005.08.011

[20] Ghamgui H, Karra-Chaâbouni M, Gargouri Y. 1-Butyl oleate synthesis by immobilized lipase from Rhizopus oryzae: A comparative study between n-hexane and solvent-free system. Enzyme and Microbial Technology. 2004;**35**:355-363. DOI: 10.1016/j.enzmictec.2004.06.002

[21] Abdul Rahman MB, Chaibakhsh N, Basri M, Raja Abdul Rahman RNZ, Salleh AB, Md Radzi S. Modelling and optimization of lipase-catalyzed synthesis of dilauryl adipate ester by response surface methodology. Journal of Chemical Technology and Biotechnology. 2008;**83**:1534-1540

[22] García J, Rodríguez F, Revenga JA. Modelling solubility of solids in supercritical fluids using response surface methodology. Journal of Chemical Technology and Biotechnology. 2000;**75**:245-251. DOI: 10.1002/(sici)1097-4660(200003)75:3<245::aid-jctb200>3.0.co;2-c

[23] Cvjetko M, Vorkapić-Furac J, Znidarsic-Plazl P. Isoamyl acetate synthesis in imidazolium-based ionic liquids using packed bed enzyme microreactor. Process Biochemistry. 2012;**47**:1344-1350. DOI: 10.1016/j.procbio.2012.04.028

[24] Romero MD, Calvo L, Alba C, Daneshfar A, Ghaziaskar HS. Enzymatic synthesis of isoamyl acetate with immobilized *Candida antarctica* lipase in n-hexane. Enzyme and Microbial Technology. 2005;**37**:42-48. DOI: 10.1016/j.enzmictec.2004.12.033

[25] Yusoff Azudin N, Mat Don M, Abd Shukor SR. Production and kinetics of isoamyl acetate from acetic anhydride using *Candida antarctica* lipase B in a solvent-free system. Chemical Engineering Transactions. 2013;**32**:1057-1062. DOI: 10.3303/CET1332177

[26] Chowdary GV, Ramesh MN, Prapulla SG. Enzymic synthesis of isoamyl isovalerate using immobilized lipase from *Rhizomucor miehei*: A multivariate analysis. Process Biochemistry. 2000;**36**:331-339. DOI: 10.1016/s0032-9592(00)00218-1

[27] Monot F, Borzeix F, Bardin M, Vandecasteele J-P. Enzymatic esterification in organic media: Role of water and organic solvent in kinetics and yield of butyrate synthesis. Applied Microbiology and Biotechnology. 1991;**35**:759-765

[28] Muthuvelayudham R, Viruthagiri T. Application of central composite design based response surface methodology in parameter optimization and on cellulase production using agricultural waste. International Journal of Chemical and Biological Engineering. 2010;**3**:97-104

[29] Ravikumar K, Pakshirajan K, Swaminathan T, Balu K. Optimization of batch process parameters using response surface methodology for dye removal by a novel adsorbent. Chemical Engineering Journal. 2005;**105**:131-138. DOI: 10.1016/j.cej.2004.10.008

Chapter 4

Obtaining Enzymatic Extract from *Pleurotus* spp. Associated with an Integrated Process for Conversion of Lignocellulosic Biomass to Bioproducts

Alma Hortensia Serafin-Muñoz,
Carlos Eduardo Molina-Guerrero,
Berenice Noriega Luna, Julio César Leal Vaca and
Aurelio Alvarez-Vargas

Additional information is available at the end of the chapter

http://dx.doi.org/10.5772/intechopen.79848

Abstract

The pretreatment of biomass has been integrated with enzyme production through the recycling of aqueous fractions. A process integrated with *Pleurotus cystidiosus* was grown, and enzymatic hydrolysis was realized. Samples of every liquid fraction from the fungal growing medium were analyzed to determine the chemical oxygen demand (OCD), glucose (Glu), xylose (Xyl), and total reducing sugars (RS). Separately, to obtain valuable polymers from this integration process, solid hemicellulose and lignin were isolated from the remaining liquid fractions through pH variation. The composition of the samples was determined using scanning electron microscopy (SEM), optical stereoscopic microscopy, and Fourier transform infrared (FTIR) spectroscopy and was compared with commercial homologs. The maximum conversion of cellulose to glucose by the obtained liquid fraction of the fungal medium was 61.3 ± 0.9% of the theoretical conversion yield of the commercial enzyme. Similarly, the conversion of hemicelluloses to xylose was 69.5 ± 1.5%. Finally, in this work, an integrated platform for cellulose, hemicellulose, lignin, enzymatic extract, and sugars production, which also significantly reduces water consumption, was proposed.

Keywords: alkaline delignification, cellulose, hemicellulose, lignin, corn straw, *Pleurotus cystidiosus*, biorefinery

1. Introduction

Pleurotus spp. is one of most extensively studied white-rot fungi for its exceptional ligninolytic properties [1]. This genus cleavages cellulose, hemicellulose, and lignin from wood, whereas brown rot fungi only cleavage cellulose and hemicellulose [2]. In basidiomycete fungi, extracellular laccases are constitutively produced in small amounts, and the lignocellulolytic enzymes are affected by many typical fermentation factors, such as medium composition, pH, temperature, aeration rate, etc. [3–5]. Mushroom survival and multiplication are related to a number of factors: chemical composition, water activity, ratio of carbon to nitrogen, minerals, surfactant, pH, moisture, sources of nitrogen, particle size, amount of inoculum, antimicrobial agents, and the presence of interactions between microorganisms. *Pleurotus* spp. is a saprophyte, and it extracts its nutrients from the substrate (grasses, wood, and agricultural residues) through its mycelium, obtaining substances necessary for its development, such as carbon, nitrogen, vitamins, and minerals. Agro-industrial waste is produced in huge amounts, and it becomes an interesting substrate for basidiomycete fungi. Many studies have been conducted to test the ability of *Pleurotus* spp. to grow on different agro-wastes, such as rice straw, wheat straw, and corn straw [6, 7].

Mexico is the third largest country in LAC in terms of the cropland area and would become a central focus of attention to produce bioproducts. It was estimated that 75.73 million tons of dry matter was generated from 20 crops in Mexico. From this biomass, 60.13 million tons corresponds to primary crop residues mainly from corn straw, sorghum straw, and wheat straw. The generation of secondary crop residues accounted for 15.60 million tons to which corncob was one the main contributors. Corn straw is the first most abundant crop residue generated in Mexico, equivalent to 66.9% of the total amount of crop residues from the cereal agro-industry [8]. The amount of corn straw reached a total of 25.1×10^6 tons, and the State of Guanajuato is the seventh largest national corn straw producer with almost 1.3×10^6 tons per year, equivalent to 5.3% of the gross production [9, 10]. This residue is mostly left to decompose or burn in situ, generating serious problems of atmospheric emissions. Nevertheless, a change in the agro-industry research topics in Mexico is emerging, where novel sustainable solutions for solid waste management are being pursued. Mexico's economic growth prospects have emphasized the importance of the development of clean technologies, in which lignocellulosic biomass conversion processes are integrated with minimum residue generation for the sustainable production of biofuels and bioproducts [11]. Three general steps are involved when transforming a lignocellulosic residue into a value-added by-product: (1) biomass pretreatment, (2) hydrolysis of the polysaccharides, and (3) fermentation of sugars. Chemical, thermochemical, and biological treatments are intended to (1) increase the formation of fermentable sugars or increase the ability to form sugars in a subsequent enzymatic saccharification step by strategically breaking the highly ordered carbohydrate structure to improve enzyme access, (2) remove or partially depolymerize lignin, (3) avoid the formation of products that are saccharification or fermentation inhibitors, and, finally, (4) be economically feasible. The alkaline/hydrogen peroxide pretreatment of corn straw (CS), also known as alkaline oxidative (AlkOx) delignification, causes the separation of biomass into its principal components: cellulose, hemicelluloses, and lignin. It also presents minimum carbohydrate degradation while enhancing further enzymatic saccharification, favoring economic feasibility. The AlkOx delignification generates products that are not limited to monosaccharides or bioethanol. They can also be commercialized as feedstocks for making composites, biodegradable packaging materials, construction materials, paper and board, easily digested cattle food, and substrates for mushroom cultivation, among other products. Alkaline pretreatment can be performed at room temperature and times ranging from seconds to days [12]. Therefore, it can reduce energy costs. Pretreatment is the key to unlocking low-cost cellulosic biomass; the pretreatment methods ought work on a widespread spectrum of feedstocks, have minimum preparatory processes, and provide a cellulosic current that can be efficiently hydrolyzed with low concentrations of enzyme [13]. In comparison with acid processing, alkaline processing features less sugar degradation and recovery or regeneration of many of the caustic salts. This leads to the reduction of costs in the process. The treatment of the water and treatment of the residues are two challenging stages for biorefineries. In this sense, as the alkaline hydrogen peroxide pretreatment of corn straw (CS) separates the biomass into cellulose, hemicelluloses, and lignin in different percentages, it enhances enzymatic cellulose hydrolysis and minimizes cellulose loss. Its products are not limited to C5/C6 sugars but can also be commercialized for polymeric applications [14, 15]. On-site production of biomass-degrading enzymes can assist in achieving full supply independence. The interest in finding new alternative treatments, such as the utilization of white-rot fungi to degrade lignocelluloses, is increasing. The white-rot fungus *Pleurotus* spp. has the ability to degrade and metabolize lignin as well as other sugars. *Pleurotus cystidiosus* is a fungus with immense biodegradation potential by the effect of its enzymes lignin oxidase, lignin peroxidase (LiP), manganese peroxidase (MnP), and laccase on the degradation of lignocellulose of corn stover. Laccases or ligninolytic peroxidases (LiP and MnP) oxidize the lignin polymer, thereby generating aromatic radicals. These evolve in different nonenzymatic reactions, including C4-ether breakdown, aromatic ring cleavage, $C\alpha$-$C\beta$ breakdown, and demethylation. The aromatic aldehyde releases from $C\alpha$-$C\beta$ breakdown of lignin or synthesized by fungi are the substrates for H_2O_2 generation by aryl-alcohol oxidase in cyclic redox reactions involving also aryl-alcohol dehydrogenases. Phenoxy radicals from C4-ether breakdown can depolymerize on the lignin polymer if they are not first reduced by oxidases to phenolic compounds. Then, lignin degradation proceeds by oxidative attack of the enzymes [16]. The objective of this work was to isolate and characterize cellulose, hemicellulose, and lignin from corn straw and the sugar concentration obtained from hydrolysis by the enzymatic cocktail from *P. cystidiosus* in an integrated scheme.

2. Materials and methods

Corn straw (*Zea mays*) was harvested at a local farm near the city of Manuel Doblado (location within the state of Guanajuato, Mexico; coordinates 20°45′08.30″N 101°56′07.31″O). We were unable to identify the crop genotype, but the most common variety used in the region is "H 368 C (1313-MAZ-556-010900/C)" developed by INIFAP (National Institute of Forest, Agricultural and Livestock Research). The planting was carried out in the month of January 2016 using common methods of 1.5 m between rows, 120 kg of seed per ha, and three irrigations in total and fertilized with phosphorus (30%) and urea with an average content of 46% nitrogen. The corn straw (CS) was first ground in a 1 HP Nogueira forage hammer mill. Granulometric separation was performed using a shaker and six sieves (Retsch, Germany) with different pore sizes (0.4, 0.5, 1, 2, 4, and 6 mm). Particles smaller than 0.4 mm were discarded [17]. The remaining material was then homogenized in a combined feedstock and stored in plastic containers at room temperature. All experiments were carried out using this feedstock. Next, it was oven-dried at 60°C for 48 h in a forced air oven. Lipids, waxes, and minerals were not removed during the procedure, for which the overall process is shown in **Figure 1**. The composition (% w/w) of the CS was determined using the Official Methods of Analysis of the Association of Official Analytical Chemists (AOAC 973.18). All chemicals used were reagent grade, purchased from Sigma-Aldrich Co. LLC (Sigma-Aldrich, St. Louis, Missouri, USA) and J.T. Baker ACS Reagent Grade (Reagents and Equipment, SA De CV, Irapuato, Gto., México). The enzyme Accellerase 1500® for the saccharification step was provided by DuPont™ Genencor® Science (DuPont, Wilmington, Delaware, USA). Commercial cellulose was also purchased from Sigma-Aldrich Co. LLC (Sigma-Aldrich, St. Louis, Missouri, USA).

2.1. AlkOx delignification

AlkOx delignification was performed using a batch system in which 60 g CS (dry matter) was soaked in 930 mL of distilled water for 30 min at 60°C in a 2000 mL flask and magnetically stirred

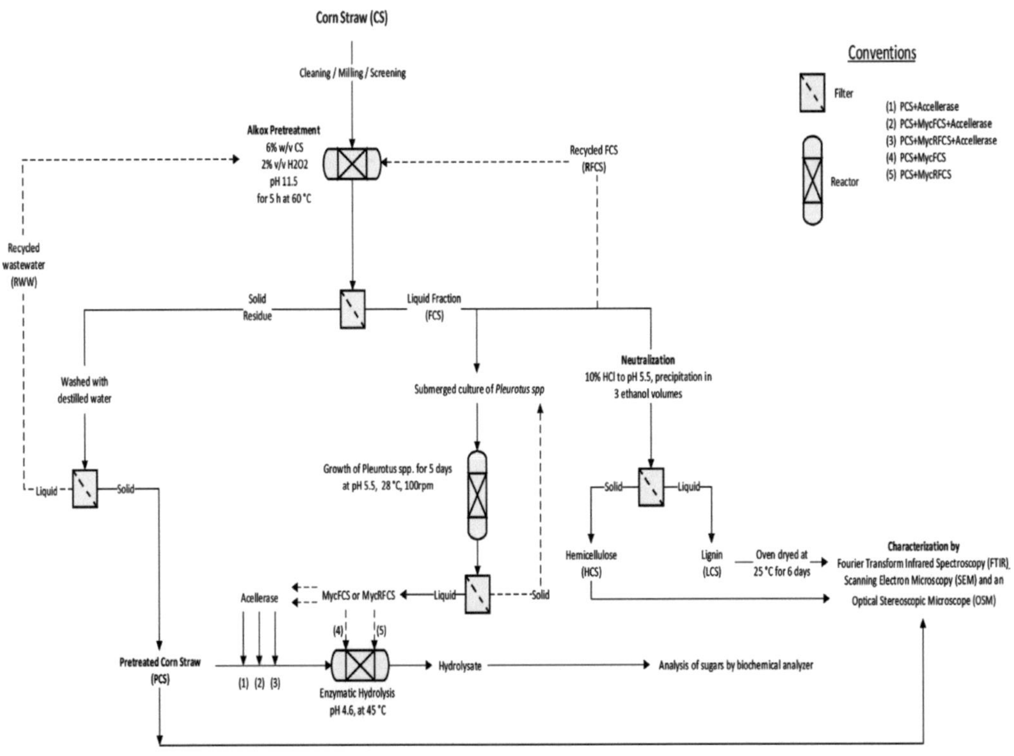

Figure 1. Global scheme for simultaneous production of glucose, xylose, cellulose, hemicellulose, lignin, and enzymes from corn straw.

at 450 rpm. Then, 35 ± 5 mL of 50% (w/v) (10 *M*) NaOH (J.T. Baker ACS Reagent Grade) solution was added until a pH of 11.5 ± 0.3 was reached. After that, 40 mL of 50% (w/v) H_2O_2 (J.T. Baker ACS Reagent Grade) was added, and the pH was again adjusted to 11.5 through the dropwise titration with 4.5 ± 1.5 mL of 5 M HCl (J.T. Baker ACS Reagent Grade) solution, giving an approximate dilution rate of 6% (w/v) for the CS. The suspension was then stirred gently for 5 h.

2.2. Isolation of bioproducts

After treatment, solid and liquid fractions were separated by filtration. The solid fraction, or pretreated corn straw (PCS), was washed five times with distilled water (EcoPura SA De CV, Santiago de Queretaro, Qro., Mexico) until the pH of the filtrate was neutral, consuming 150–250 mL of water per gram of straw. Then, PCS was dried at 50°C for 24 h. The solid fractions after AlkOx treatment were mainly formed of crude cellulose. The water was recovered (RWW) and reintegrated in subsequent processes, as depicted in **Figure 1**. The liquid fraction (FCS) was kept and stored at 4°C, until further use. Hemicellulose was obtained by precipitation. First, the FCS was adjusted to pH 5.5 with 5 M HCl. The solubilized hemicelluloses were precipitated by adding anhydrous ethanol at a 3:1 (v/v) ratio. Then, the sample was centrifuged at 7000 rpm for 10 min, and, finally, the solid was dried in a convection oven at 60°C for 8 h. Lignin was obtained by drying the remaining liquid fraction for 6 days at 25°C.

2.3. Characterization

Infrared spectrum of PCS, hemicellulose, and lignin were obtained with a PerkinElmer Spectrum One FTIR spectrophotometer (Perkin Elmer de Mexico, SA, Mexico, DF) using the attenuated total reflection (ATR) device. Thirty-two scans were taken from each sample and recorded from 4000 to 650 cm^{-1} at a resolution of 2 cm^{-1} in transmission mode. Automatic ATR, baseline correction, and normalization with a factor of 0.5 were performed on each spectrum. For the scanning electron microscopy (SEM) imaging, the untreated (UCS) and PCS were sputter coated with gold for 100 s. The coated samples were observed using a JEOL JSM-35CF microscope (JEOL de Mexico SA De CV, Mexico, DF) operated at 15 kV. Stereoscopic images were taken with an optical stereoscopic microscope (OSM); the model was Euromex Novex trinocular zoom stereo microscope, model RZT-SF 65.560 (Westek, SA DE CV, Guadalajara Jal., Mexico).

2.4. Submerged culture of mycelium

The white-rot fungus was isolated from local wood residues, which was identified as *Pleurotus cystidiosus* strain P-24 (GenBank FJ379283.1) by the Instituto Potosino de Investigacion Científica y Tecnológica (IPICYT, San Luis Potosí, SLP, Mexico). The fungus was cultured in 125 mL Erlenmeyer flasks containing 50 mL of FCS or recycled FCS (RFCS) at pH 5.5 after inoculation with two 2 mm spheres sampled from a plate with PDA on which the fungal mycelia had been cultured. The flasks were set in an orbital shaker incubator at 28°C and 100 rpm for 5 days. Then, filtrates were obtained (MycFCS and Myc**R**FCS) from the cultures.

2.5. Enzymatic hydrolysis

The enzyme Accellerase 1500™ for the saccharification step was provided by DuPont™ Genencor® Science (DuPont, Wilmington, Delaware, USA) [18]. The filter paper activity was determined by the standard procedures recommended by the National Renewable Energy Laboratory (NREL/TP-510-42,628). The hydrolysis of the PCS was performed in a thermomixer (Eppendorf Thermomixer Comfort, Surtidor Quimico del Centro SA De CV, Santiago de Queretaro Qro., Mexico) at 40°C and 100 rpm [19]. The samples were placed in 1.5 mL microcentrifuge tubes containing 1% w/v of dry PCS with either distilled water or MycFCS or Myc**R**FCS cultures. The experiments were run as follows: PCS + MycFCS, PCS + MycRFCS, PCS + MycFCS + Accellerase, PCS + MycRFCS + Accellerase, and PCS + Accellerase. Each experiment was stopped at different times (24, 48, and 120 h), alone or mixed with an enzyme loading of 16 FPU g^{-1} of PCS at pH of 5.0 (0.1 mol L^{-1} sodium acetate buffer), with sodium azide (NaN_3, 0.01% (w/v)) to prevent microbial growth.

2.6. Analysis of the liquid phases

Hydrolysate was separated from the solid residue by centrifugation at 8608.6 g-force (Eppendorf rotor FA-45-18-11, Equipar SA De CV, Ciudad de Mexico, Mexico) for 10 min. The total reducing sugars (RS) were measured using the Miller method [20]. The measurement of glucose (Glu) and xylose (Xyl) was accomplished in the biochemical analyzer 2700 SELECT from YSI (YSI Life Sciences, Yellow Springs, Ohio, USA) using the Xylose Software. The OCD analyses were run according to the standard method for the analysis of potable and residual water [21]. All these experiments were conducted in triplicate.

2.7. Data analysis

Statistical analysis was developed in order to determine the statistical differences between treatments against time using the software SAS JMP version 13.1.0 (SAS Inc., Cary, North Carolina). First, a one-way analysis of variance (ANOVA) was carried out to determine if the results obtained using different treatments were significantly different. Then, if the ANOVA confirmed the existence of a significant difference ($p < 0.05$), a post hoc analysis (Tukey's HSD) was used to determine between which values the difference was significant, considering a level of significance of 0.05.

3. Results and discussion

The samples of CS had the following composition (% w/w): 43.5% cellulose, 35.8% hemicellulose, 11.1% lignin, and 9.2% other components on a dry weight basis (**Table 1**). The lignin content of CS was slightly lower than in other reported works [22–24]. However, within the range, it coincides with other published works [25, 26]. Considerable variation in feedstock composition may occur because the corn straw is grown in different environments. In addition, there can be genotypic differences. Feedstock composition varies with crop maturity.

Components	Percentage (%)
Dry matter	90.2 ± 3.3
Cellulose	43.5 ± 1.6
Hemicellulose	35.8 ± 1.2
Lignin	11.1 ± 1.8
Ash	9.2 ± 2.0
Calcium	0.3 ± 0.1
Phosphorus	0.05 ± 0.04

Bromatological analysis of WS realized by the University Center for Biological and Agricultural Sciences (CUCBA) at the University of Guadalajara, Mexico, using the official methods of National Renewable Energy Laboratory (NREL).

Table 1. Chemical components of corn straw (CS) from Manuel Doblado, Gto., Mexico (20°43′42″N 101°56′57″O).

Maximum lignocellulosic yield is obtained when corn is harvested at physiological maturity. When the corn straw is harvested before and after physiological maturity, the compositions of lignocellulose are found to be distinct.

3.1. AlkOx delignification

Cellulose, hemicellulose, and lignin contents after the pretreatment are presented in **Table 2**. Compositional analysis of the PCS showed that it had 75.1% cellulose, 5.9% hemicelluloses, and 0.6% lignin. Under these conditions, 83.5% of the hemicellulose and 93.4% of the lignin were removed after pretreatment. Previous reports about the alkaline oxidative pretreatment of corn straw for bioproduct production are similar [26–30]. Comparing the results with other studies for the delignification of CS, the lignin content was slightly lower than that obtained by us [27]. Gould [31] demonstrated the use of H_2O_2 for delignification with a maximum pH of 11.5. The porosity of the lignocellulosic materials increases with the removal of the cross-link dilute NaOH treatment of lignocellulosic materials causing swelling, leading to an increase in internal surface area, a decrease in the degree of polymerization, a decrease in crystallinity, a separation of structural linkages between lignin and carbohydrates, and a disruption of the lignin structure [32]. On the other hand, the pretreatment of CS with hydrogen peroxide enhanced its susceptibility to enzymatic hydrolysis. As seen in **Table 2**, Xyl was the most abundant sugar present, suggesting that some hemicellulose degradation had occurred. This is due to the partial hemicellulose degradation, and its magnitude increases proportionally with the temperature, pH, and reaction time period. On the other hand, during this process, the reuse of wastewater from washing of the PCS resulted in low water consumption (**Figure 1**). The quantity of water used to wash was about 1.5 L per 50 g of the PCS (it should be carried out at pH 6.0 ± 1.5), which was reintegrated to 100% into the process. Considering the abovementioned, 30,000 L of water could be wasted per 1 ton of the PCS into the process. It is noteworthy that the moisture content of PCS was between 8 and 12%. This AlkOx pretreatment, as proposed in this study, has several potential advantages over the alkaline wash step, because it presents a higher amount of sugars, higher solid loadings can be treated, and

Pretreated wheat straw (PWS)	Cellulose (%)	Hemicellulose (%)	Lignin (%)	RS (%)	Glu (%)	Xyl (%)
6 mm screen size WS	75.1 ± 2.1	5.9 ± 1.1	0.6 ± 0.1	0.7 ± 0.08	0.2 ± 0.02	0.4 ± 0.05
Component removed (CR)	—	83 ± 1.1	93 ± 1.3	—	—	—
Kristensen [58]	75	90	100			
Asghar [29]	83	90	81	—	—	—

Table 2. Compositional analysis of the pretreated corn straw (solid fraction), total reducing sugars (RS), glucose (Glu), and xylose (Xyl) released (liquid fraction) and references for component removal.

a significantly lower amount of water is required. All of these aspects have a direct impact on the manufacturing costs and energy requirements of the process [33].

3.2. Characterization

Results from FTIR spectroscopy, SEM, and OSM were used to compare bioproducts (cellulose, hemicellulose, and lignin) obtained, as well as information on the effect of the pretreatment on the structure and possible disruption of the cell wall (**Figures 2–5**). The FTIR spectrum of the commercial cellulose and the PCS were similar (**Figure 2a** and **b**). The peaks at 1165 at 1059 cm^{-1} were assigned to the linkage present in the cellulose and can be associated to β(1–3)-polysaccharide, which is a strong signal characterizing a high cellulose composition [34]. The transmittance at 1429, 1363, 1317, 1161, 1051, and 896 cm^{-1} (**Figure 2a** and **b**) is a typical pure cellulose that are associated with CH_2 in-plane bending vibrations, with C—O stretching, and with C—H ring in-plane bending vibrations [35–37]. The absence of a band

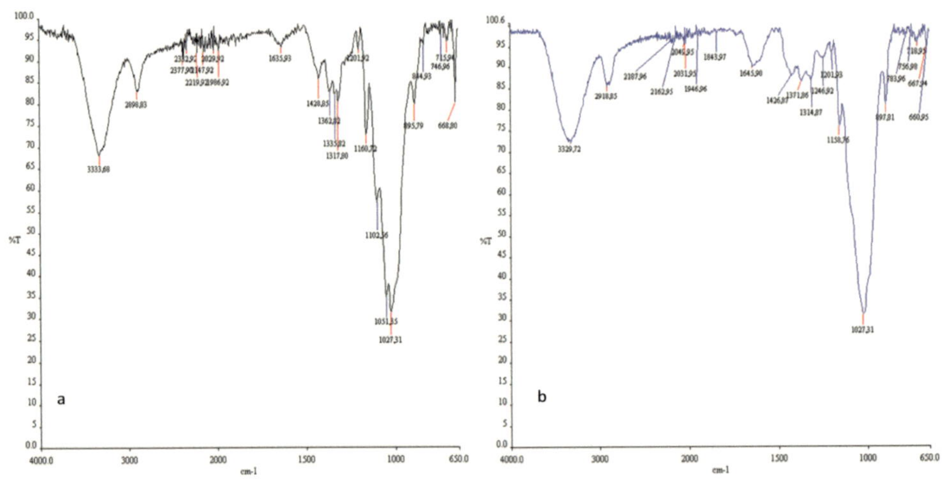

Figure 2. Attenuated total reflectance-Fourier transform infrared (ATR-FTIR) spectra of (a) commercial cellulose and (b) pretreated corn straw.

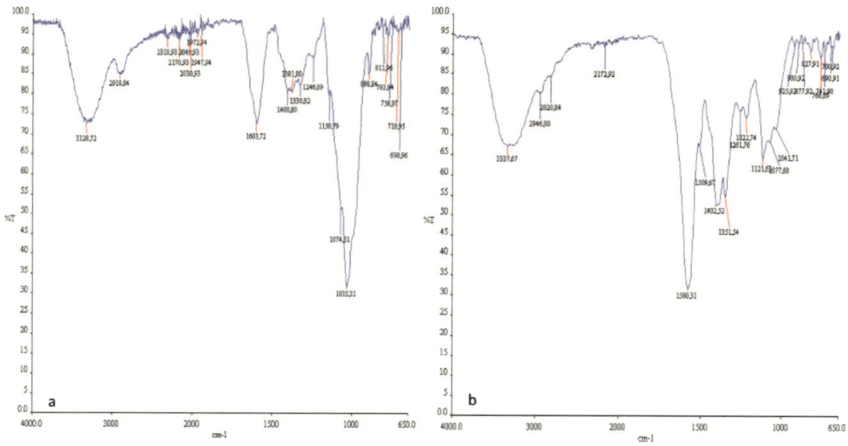

Figure 3. Attenuated total reflectance-Fourier transform infrared (ATR-FTIR) spectra of (a) hemicellulose and (b) lignin from alkali pretreatment of corn straw.

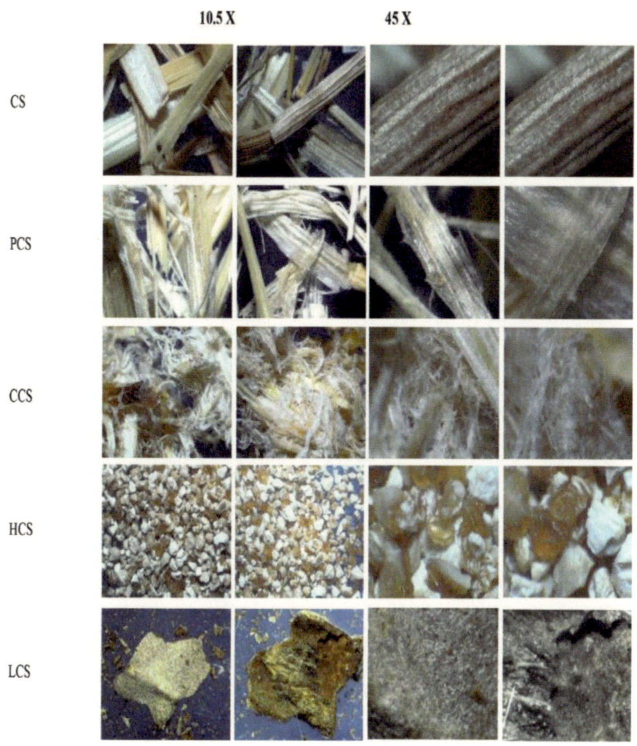

Figure 4. Images obtained by optical stereoscopic microscopy of the untreated (CS), pretreated (PCS), cellulose (CCS), hemicellulose (HCS), and lignin (LCS) from corn straw.

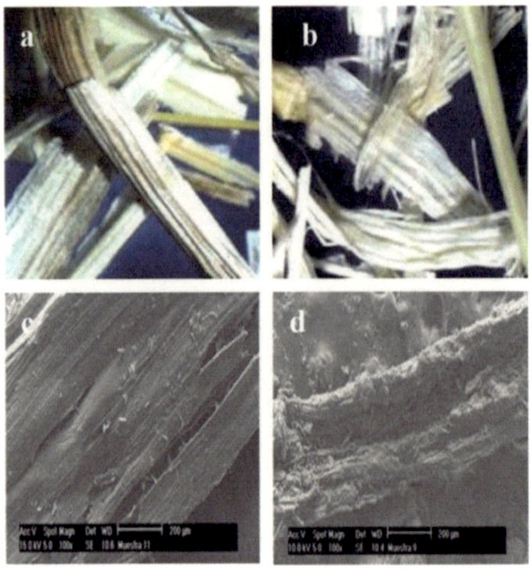

Figure 5. Optical stereoscopic microscopy (a, b at 10.5×) and scanning electron microscopy (c, d at 100×) images of the untreated (a, c), and pretreated corn straw (b, d).

at 1731 cm^{-1} indicates cleavage of the acetyl and uronic ester groups from the hemicelluloses, while the absence of the absorption band at 1750–1700 cm^{-1} region and the peak relating to lignin aromatic ring vibrations at 1513 cm^{-1} revealed that lignin was removed during the alkaline peroxide process [34]. These results corroborate the ones obtained for the percentage composition of the PCS obtained in the present work. **Figure 3** shows the FTIR spectrum of the hemicelluloses and lignin. The absorbance at 2919, 1408, 1382, 1247, 1075, 1036, 985, and 899 cm^{-1} is associated with hemicelluloses, in which 1039 cm^{-1} is typical of arabinoxylans (**Figure 3a**). The FTIR spectrum of lignin is shown in **Figure 3b**. The C—H stretching vibration gives signals at 2919–2820 cm^{-1} [38]. It is noteworthy that the structure of lignin varies with different types of agricultural crops [39, 40]. In the case of the spectrum, an intensive band at 1580 cm^{-1} was observed, while it was a weak peak at 1509 cm^{-1}, which are associated with quadrant ring stretching and semicircle ring stretching (aromatic lignin) [34]. In addition, the absorption at 3334–3328 cm^{-1} is attributed to the stretching of —OH groups that may include absorbed water [27]. The C—H stretching vibration gives signals at 2919–2820 cm^{-1}. The band at 1646–1636 cm^{-1} is due to the bending mode of absorbed water [27]. Specific band in the 1200–1000 cm^{-1} regions is dominated by ring vibrations overlapped with stretching vibrations of side groups (C—OH) and the glycosidic bond vibration (C—O—C) [34]. The high absorbance at 1351 cm^{-1} arises from the C—C and C—O skeletal vibrations [34]. **Figure 4** shows the different structures of the products obtained from the corn straw by OSM. PC is clearly observed that color changes from brown to slightly white, as well as structural deformation of fiber cells and particle fragmentation by pretreatment, indicate lignin removal. In the cellulose image, the fibrous structure is observed, unlike the hemicellulose. Hemicellulose shows a white to yellow and faint brown to brown color. Other works have mentioned that, at high NaOH loadings (above 260 mg NaOH/g dry straw), there is a diminishing effect on lignin removal as the NaOH loading is increased [41], so that the pretreated solids range from dark brown to white as more lignin is removed.

Figure 5 shows the characterization of PCS and UCS by SEM and OSM. Lignin shows a light brown color, and it presents a rubberlike appearance. These characteristics are in accordance with the standards described by commercial houses [42]. As seen in **Figure 5a** and **c**, the OSM image and SEM micrograph, the anatomy of the UCS and PCS is easily recognizable, the thick-walled fiber of the straw wall can be seen, as well as it is largely made of wax. The most evident effect of the pretreatment is the separation of individual fibers of the CS. On the other hand, the pretreated material is quite heterogeneous and contains pieces of different sizes (<1 cm) (**Figure 5b**). Cracks and holes were seen in the structure of the corn straw (**Figure 5d**). This shows that it is easily digestible by enzymes [16, 17].

3.3. Submerged culture of the mycelium

Figure 6 shows the time course of the biomass of the mushroom strain *(P. cystidiosus)*, and the sugar concentration is reported for the onset of mushroom growth. The RS concentrations from FCS and **RFC** were 6.9 ± 0.7 and 7.3 ± 1.8 g/L. Percentage contents of xylose were 68.1 and 73.4%, respectively. This bioprocess presented an interesting behavior. A diauxic growth was observed, an inflection in the growth curve or even a decline in biomass occurs at 48 h. Under these conditions, 0.45% (w/v) of the amount of biomass was increased to 0.69% (w/v) at 72 h, which represents 53% (w/w). The presence of a preferential carbon substrate may inhibit the synthesis of the enzymatic system involved in the uptake and metabolism of a second carbon substrate. The *P. cystidiosus* culture may metabolize xylose preferentially in a mixed substrate medium containing xylose and other sugars. The product of xylose metabolism would be

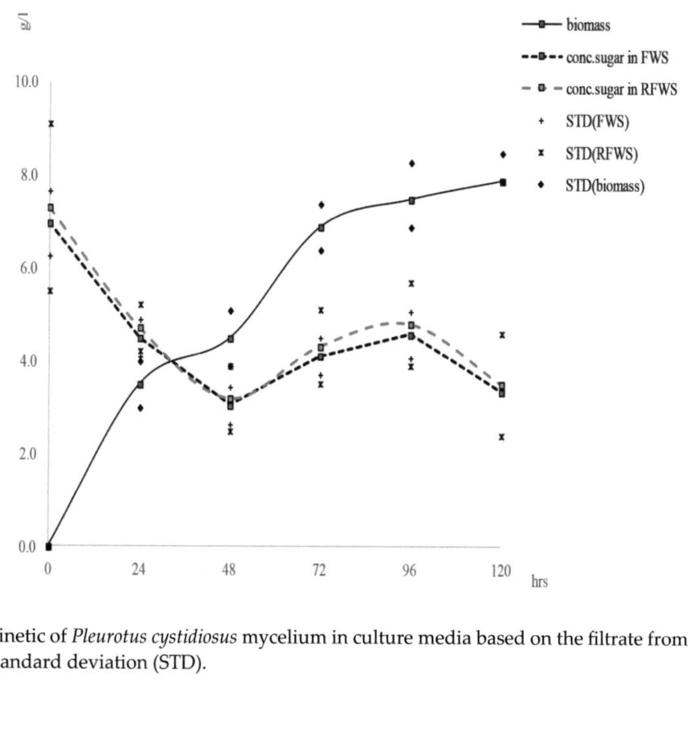

Figure 6. Growth kinetic of *Pleurotus cystidiosus* mycelium in culture media based on the filtrate from the process of PCS (FCS and **R**FCS). Standard deviation (STD).

utilized as a carbon source only when the principal one is exhausted. There are few reported works on the growth of *P. cystidiosus*; these mention fungi grown on a medium containing glucose [43, 44]. However, xylose is shown as a good carbon source by several works [45]. The RS concentration after 96 h decreased to 73%, coinciding with the poor growth of biomass. In this context, **Figure 7** shows the relation between OCD and growth of biomass. The OCD was reduced to 75%, which is congruent with the increase in biomass. In the stationary phase (72 h), oxygen consumption dropped drastically, indicating a decrease in cellular aerobic metabolism, which may be associated with lack of nutrients in the medium or loss of cellular viability. In the same context, in spite of the numerous potential applications of the *Pleurotus* spp. enzymatic system, little is known about the production of enzymes, in which dissolved oxygen concentration, among other parameters, can be strictly controlled [46].

3.4. Enzymatic hydrolysis

Figure 8a shows the conversion of cellulose to glucose due to hydrolysis using MycFCS and Myc**R**FCS, alone or mixed with Accellerase 1500. **Figure 8b** shows the conversion of hemicellulose to xylose. The maximum sugar concentration was obtained by PCS + Myc**R**FCS + Accellerase for 120 h, with 6.8 g/L of RS, which included 4.6 g/L Glu and 0.41 g/L Xyl, representing a conversion of cellulose to glucose (CCTG) about 61.3% and a conversion of hemicellulose to xylose (CHTX) around 69.5%. When only the MycFCS medium was used, the optimal result occurred when using PCS + Myc**R**FCS-48 h with a conversion and a net RS increment of 2.9 g/L, which included 2.6 g/L Glu and 0.30 g/L Xyl, representing a CCTG of 34.7% and CHTX of 50.8%. In the case of PCS + MycFCS, the results obtained for CCTG at 48 and at 120 h were 29.7 and 30.5%, respectively. Within the same case, for CHTX were 49.9 and 49.1%. The lowest concentrations were for the experiment PCS + MycFCS for 24 h, 12.2% of CCTG, and 20.1% of CHTX. In all cases, adding Accellerase resulted in slightly higher (CCTG (28.1–61.3%) and CHTX (30–69.5%)); however, the concentrations were lower when Accellerase was used without any MycFCS or Myc**R**FCS (CCTG (26.3–49.1%) and CHTX (15.2–29.7%)). Therefore, a synergistic effect was

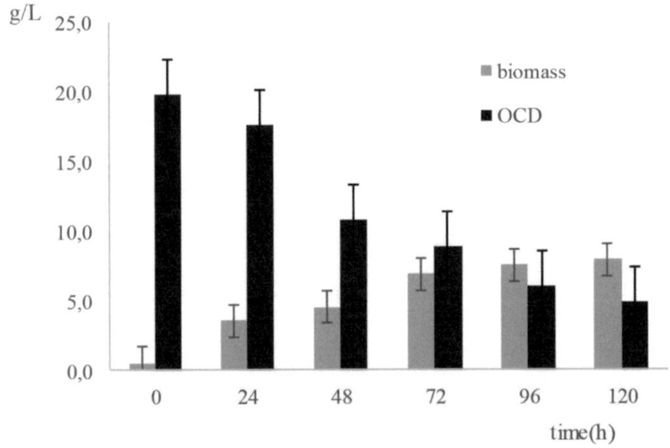

Figure 7. The relation between chemical oxygen demand and growth of biomass.

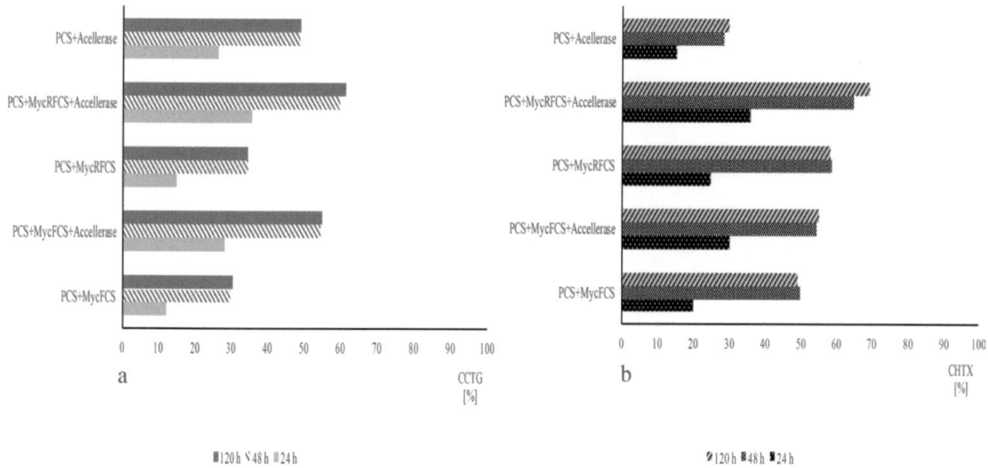

Figure 8. (a) Conversion of cellulose to glucose due to hydrolysis using the filtrates from the growth of the fungus (MycFCS and Myc**R**FCS), alone or mixed with Accellerase 1500. (b) Conversion of hemicellulose to xylose due to hydrolysis using the filtrates from the growth of the fungus (MycFCS and Myc**R**FCS), alone or mixed with Accellerase 1500.

observed among the mixtures. Some studies have reported the conversion of cellulose to glucose above 80%; in our case the results were lower than expected [47, 48]. This might be due to the amounts of the substrate released to the MycFCS or Myc**R**FCS, inhibiting the enzymatic hydrolysis process. On the other hand, it is documented that all the species of *Pleurotus* produced laccase, manganese peroxidase, and aryl-alcohol oxidase activity [49–51]. Then, this may have an effect of competitive inhibition with enzyme complex from Accellerase, which contains multiple enzyme activities: exoglucanase, endoglucanase, hemicellulase, and beta-glucosidase. However, one of the important contributions of this work is that *P. cystidiosus* cultured in lignocellulosic substrates in FCS or RFCS was found to secrete a range of important degradative enzymes, such as xylanase (CHTX (20.1–50.8%)). Moreover, it should be noted that the MycFCS or Myc**R**FCS comes from the mycelial stage of the fungus. Otherwise, Ellisahvilli observed that laccase activity of *Pleurotus* was high during the colonization stage and declined during the first primordial formation and fruiting stage [52]. Similar result was reported by Malarczyk in *P. cystidiosus* where laccase was active during mycelial growth on solid saw dust [53]. The enzymes associated with lignin-degrading ability of white-rot fungi are lignin peroxidase, manganese peroxidase, laccase, and xylanases [54]. Although it is known that white-rot fungi are known to employ a variety of extracellular ligninolytic enzymes, in the case of the *P. cystidiosus*, there are few reported works, and little attention has been given to the evaluation of the hydrolytic system of this fungus [55–58].

3.5. Data analysis

The concentration range of RS was 1.33–6.9 g/L. The arithmetic average and relative standard deviation of all analyses are performed; sample size in $n = 45$ was 4.5 ± 1.7 g/L. Similar results are found in treatments PCS + MycFCS (Treatment 1) and PCS + M_{YC}**R**FCS (Treatment 3),

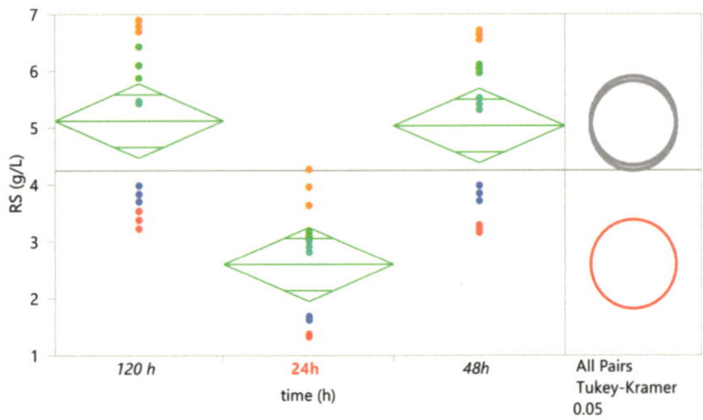

Figure 9. Univariant analysis and comparison means (Tukey's HSD) for RS from PCS by the effect of enzymatic hydrolysis time.

Source	Degrees of freedom	Sum of squares	Mean square	F ratio	Prob > F
Treatment	4	63.6	15.9	10.1	<.0001
Time (h)	2	61.356	30.678	19.6472	<.0001

Table 3. Analysis of variance for the statistical differences between RS from PCS by different enzymatic hydrolysis treatments and time (h) using the software SAS JMP version 13.1.0 (SAS Inc., North Carolina).

Level	- Level	Difference	Std Err Dif	Lower	Upper	p-Value	
PCS+MycRFCS+Accellerase	PCS+MycFCS	3.14	0.59	1.45	4.84	<.0001	
PCS+MycRFCS+Accellerase	PCS+MycRFCS	2.68	0.59	0.99	4.38	0.00	
PCS+MycFCS+Accellerase	PCS+MycFCS	2.45	0.59	0.75	4.14	0.00	
PCS+MycFCS+Accellerase	PCS+MycRFCS	1.98	0.59	0.29	3.68	0.01	
PCS+Acellerase	PCS+MycFCS	1.94	0.59	0.24	3.63	0.02	
PCS+Acellerase	PCS+MycRFCS	1.48	0.59	-0.22	3.17	0.11	
PCS+MycRFCS+Accellerase	PCS+Acellerase	1.20	0.59	-0.49	2.90	0.27	
PCS+MycRFCS+Accellerase	PCS+MycFCS+ Accellerase	0.70	0.59	-1.00	2.39	0.76	
PCS+MycFCS+Accellerase	PCS+Acellerase	0.51	0.59	-1.19	2.20	0.91	
PCS+MycRFCS	PCS+MycFCS	0.46	0.59	-1.23	2.16	0.94	
120 h	24h	2.52	0.46	1.41	3.63	<.0001	
48h	24h	2.43	0.46	1.32	3.54	<.0001	
120 h	48h	0.09	0.46	-1.02	1.20	0.98	

Table 4. Ordered difference report between RS from PCS by different enzymatic hydrolysis treatments and time (h).

PCS + MycFCS + Accellerase (Treatment 2), and PCS + MycRFCS + Accellerase (Treatment 4). The ANOVA applied to PCS + Accellerase (Treatment 5) shows that significant statistical differences (SSD) exist between all treatments and are possible to suggest that the experiment could be performed over a long time with respect to orders. **Figure 9** shows that results were very similar to each other where there are no SSD between the results obtained at 48 and 120 h, but all the treatments showed SSD with respect to 24 h. The results suggest that the process is completed at 48 h. **Tables 3** and **4** show the analysis of variance for the statistical differences between RS from PCS by different enzymatic hydrolysis treatments and time (h), and the ordered differences report hydrolysis treatments and time (h), respectively.

4. Conclusions

The integrated scheme with the use of the filtrate from the pretreatment of the CS and the growth conditions of *P. cystidiosus* studies showed a great potential for the production of lignocellulolytic enzymes and application of crude enzymatic extract by the fungus. Replacing commercial enzymes with locally produced enzymes may reduce the cost of enzymatic saccharification, (e.g., the approximate cost of commercial cellulose from Sigma-Aldrich is about $110 USD of 50 mL of aqueous solution (≥700 units/g)), which may translate to high-impact savings in the production cost of fermentable sugars. From a perspective of clean and sustainable technologies, a very important contribution of this work is the reuse of effluents from the treatment of the delignification of lignocellulosic waste to integrate them within the same chain of processes. This way to reduce considerably the use of resources like water and raw material in parallel reduces cost of production. In addition, the utilization of agro-industrial wastes as feedstock in the production of enzymes is more economical and profitable. The high cost of raw materials in the production of enzymes is around 40-60%. Therefore, agro-industrial wastes strategically represent low-cost raw materials, due to their high biodegradability and rich in carbon, to be used as substrates in the chain of production of enzymes. Another important contribution is about the fate of the corn straw. One of the environmental problems in several regions of Mexico is because the straw is burned by the farmers. Therefore, using it as a raw material for obtaining high-value-added bioproducts contributes to improve air quality, environment, and quality of life. Finally, these results can be part and contribute to the development of a biorefinery which can be applied in the treatment of lignocellulosic waste. Undoubtedly, in Mexico, the biorefineries represent great opportunities to harness the economic benefit of agro-industrial waste and to develop even more efficient and sustainable systems.

Acknowledgements

The authors would like to thank FINNOVATEG-CFINN0186 (SICES Project) and to the Directorate for Research Support and Postgraduate Programs at the University of Guanajuato for their support in the translation and editing of the English language version of this article.

List of abbreviations

CS	corn straw
UCS	untreated corn straw
PCS	pretreated corn straw
FCS	liquid fractions from PCS
RFCS	recycled FCS
MycFCS	filtrates of cultures from FCS + mycelium
MycRFCS	filtrates of cultures from RFCS + mycelium
SSD	significant statistical difference
STD	standard deviation

Author details

Alma Hortensia Serafin-Muñoz[1*], Carlos Eduardo Molina-Guerrero[2], Berenice Noriega Luna[1], Julio César Leal Vaca[1] and Aurelio Alvarez-Vargas[1]

*Address all correspondence to: sermuah@ugto.mx

1 Engineering Division, University of Guanajuato, Guanajuato, Mexico

2 Faculty of Chemical Sciences, Autonomous University of Chihuahua, Chihuahua, Mexico

References

[1] Economou CN, Diamantopoulou PA, Philippoussis AN. Valorization of spent oyster mushroom substrate and laccase recovery through successive solid state cultivation of *Pleurotus*, Ganoderma, and Lentinula strains. Applied Microbiology and Biotechnology. 2017;**101**(12):5213-5222

[2] Machado DL et al. Adsorption characteristics of cellulase and beta-glucosidase on Avicel, pretreated sugarcane bagasse, and lignin. Biotechnology and Applied Biochemistry. 2015;**62**(5):681-689

[3] Velioglu Z, Ozturk Urek R. Optimization of cultural conditions for biosurfactant production by *Pleurotus* djamor in solid state fermentation. Journal of Bioscience and Bioengineering. 2015;**120**(5):526-531

[4] Cilerdzic J et al. Potential of selected fungal species to degrade wheat straw, the most abundant plant raw material in Europe. BMC Plant Biology. 2017;**17**(Suppl 2):249

[5] Fokina O et al. Selective natural induction of laccases in *Pleurotus* sajor-caju, suitable for application at a biofuel cell cathode at neutral pH. Bioresource Technology. 2016;**218**:455-462

[6] Alananbeh KM, Bouqellah NA, Al Kaff NS. Cultivation of oyster mushroom *Pleurotus* ostreatus on date-palm leaves mixed with other agro-wastes in Saudi Arabia. Saudi Journal of Biological Sciences. 2014;**21**(6):616-625

[7] Serafin Muñoz AH, Molina Guerrero CE, Gutierrez Ortega NL, et al. Characterization and integrated process of pretreatment and enzymatic hydrolysis of corn straw. Waste Biomass Valor. 2018. https://doi.org/10.1007/s12649-018-0218-9

[8] Hellin J et al. Maize Stover use and sustainable crop production in mixed crop–livestock systems in Mexico. Field Crops Research. 2013;**153**:12-21

[9] Nacional S.D.I.A.Y.P. Cierre de la producción agrícola por estado. 2015. http://www.siap.gob.mx/cierre-de-la-produccion-agricola-por-estado. [Accessed: November 2015]

[10] Guzmán Soria E et al. Análisis de los costos de producción de maíz en la Región Bajío de Guanajuato. Análisis Económico. 2014;**XXIX**(70):145-155

[11] Marousek J et al. Techno-economic assessment of processing the cellulose casings waste. Clean Technologies and Environmental Policy. 2015;**17**:2441-2446

[12] Alvira P et al. Pretreatment technologies for an efficient bioethanol production process based on enzymatic hydrolysis: A review. Bioresource Technology. 2010;**101**(13):4851-4861

[13] Chandra RP et al. Substrate pretreatment: The key to effective enzymatic hydrolysis of lignocellulosics. Advances in Biochemical Engineering/Biotechnology. 2007;**108**:67-93

[14] Tang Y et al. Enhancement of fermentable sugar yield by competitive adsorption of non-enzymatic substances from yeast and cellulase on lignin. BMC Biotechnology. 2014;**14**:21

[15] Cynthia AA et al. Reduction in personal exposures to particulate matter and carbon monoxide as a result of the installation of a Patsari improved cook stove in Michoacan Mexico. Indoor Air. 2008;**18**(2):93-105

[16] Martinez AT et al. Biodegradation of lignocellulosics: Microbial, chemical, and enzymatic aspects of the fungal attack of lignin. International Microbiology. 2005;**8**(3):195-204

[17] Moniz P et al. Characterisation and hydrothermal processing of corn straw towards the selective fractionation of hemicelluloses. Industrial Crops and Products. 2013;**50**:145-153

[18] ACCELLERASE® 1500. Genencor, Accellerase®, TRIO TM and Danisco are trademarks or registered trademarks of E. I. du Pont de Nemours and company or its affiliates. Danisco US Inc. 2011. GENENCOR and ACCELLERASE are registered trademarks of Danisco US Inc. or its affiliated companies; 2011. pp. 1-4

[19] Chen BY et al. Use of different alkaline pretreatments and enzyme models to improve low-cost cellulosic biomass conversion. Biomass and Bioenergy. 2012;**39**:182-191

[20] Miller GL. Use of dinitrosalicylic acid reagent for determination of reducing sugar. Analytical Chemistry. 1959;(3):**31**

[21] Norma oficial mexicana nom-127-ssa1-1994. Salud ambiental, agua para uso y consumo humano-limites permisibles de calidad y tratamientos a que debe someterse el agua para su potabilizacion. Seismological Society of America. 1994. http://www.salud.gob.mx/unidades/cdi/nom/127ssa14.html

[22] Zhang CZ et al. Multilevel composition fractionation process for high-value utilization of wheat straw cellulose. Biotechnology for Biofuels. 2014;**137**(7):1-12

[23] Skaggs L et al. Waste-to-energy biofuel production potential for selected feedstocks in the conterminous United States. Renewable and Sustainable Energy Reviews. 2018;**3**(82): 2640-2265

[24] Del Rio JC. Structural characterization of wheat straw lignin as revealed by analytical pyrolysis, 2D-NMR, and reductive cleavage methods. Journal of Agricultural and Food Chemistry. 2012;**60**(23):5922-5935

[25] Mubeen TSKAU. Wheat Straw: A pragmatic overview. Journal of Biological Sciences. 2012;**4**(6):673-675

[26] Suna RCT et al. Physico-chemical and structural characterization of hemicelluloses from wheat straw by alkaline peroxide extraction. Polymer. 2000;**41**:2647-2656

[27] Kristensen JB et al. Cell-wall structural changes in wheat straw pretreated for bioethanol production. Biotechnology for Biofuels. 2008;**1**(1):5

[28] Hansen MA et al. Pretreatment and enzymatic hydrolysis of wheat straw (*Triticum aestivum* L)—The impact of lignin relocation and plant tissues on enzymatic accessibility. Bioresource Technology. 2011;**102**(3):2804-2811

[29] Asghar U et al. Effect of alkaline pretreatment on delignification of wheat straw. Natural Product Research. 2015;**29**(2):125-131

[30] Wang Q et al. Cell wall disruption in low temperature NaOH/urea solution and its potential application in lignocellulose pretreatment. Cellulose. 2015;**22**:3559-3568

[31] Gould JM. Alkaline peroxide delignification of agricultural residues to enhance enzymatic saccharification. Biotechnology and Bioengineering. 1984;**26**(1):46-52

[32] Ibbett R et al. Structural reorganisation of cellulose fibrils in hydrothermally deconstructed lignocellulosic biomass and relationships with enzyme digestibility. Biotechnology for Biofuels. 2013;**6**(1):33

[33] Stoklosa RJ et al. Techno-economic comparison of centralized versus decentralized biorefineries for two alkaline pretreatment processes. Bioresource Technology. 2017;**226** (Suppl. C):9-17

[34] Adapa PK, Schonenau PL, Thomas C, Dumonceaux T, et al. Quantitative analysis of lignocellulosic components of Non-treated and steam exploded barley, canola, oat and wheat straw using fourier transform infrared spectroscopy. Faculty Research & Creative Activity. 2011;107. http://thekeep.eiu.edu/bio_fac/107

[35] Alemdar A, Sain M. Isolation and characterization of nanofibers from agricultural residues: Wheat straw and soy hulls. Bioresource Technology. 2008;**99**(6):1664-1671

[36] Kaushik A, Singh M. Isolation and characterization of cellulose nanofibrils from wheat straw using steam explosion coupled with high shear homogenization. Carbohydrate Research. 2011;**346**(1):76-85

[37] Montano-Leyva B et al. Preparation and characterization of durum wheat (*Triticum durum*) straw cellulose nanofibers by electrospinning. Journal of Agricultural and Food Chemistry. 2011;**59**(3):870-875

[38] Kirtania K et al. In situ synchrotron IR study relating temperature and heating rate to surface functional group changes in biomass. Bioresource Technology. 2014;**151**:36-42

[39] Yang H et al. Comparative study of lignin characteristics from wheat straw obtained by soda-AQ and Kraft pretreatment and effect on the following enzymatic hydrolysis process. Bioresource Technology. 2016;**207**:361-369

[40] Zhang LH et al. Effect of steam explosion on biodegradation of lignin in wheat straw. Bioresource Technology. 2008;**99**(17):8512-8515

[41] Karp E et al. Alkaline pretreatment of corn stover: Bench-scale fractionation and stream characterization. ACS Sustainable Chemistry & Engineering. 2014;**2**:1481-1491

[42] http://www.sigmaaldrich.com/Graphics/COfAInfo/SigmaSAPQM/SPEC/31/310697/310697-BulkAldrich

[43] Panda BC et al. Heteroglycan of an edible mushroom *Pleurotus cystidiosus*: Structural characterization and study of biological activities. International Journal of Biological Macromolecules. 2017;**95**:833-842

[44] Atri N et al. Nutritional and neutraceutical composition of five wild culinary-medicinal species of genus *Pleurotus* (higher Basidiomycetes) from Northwest India. International Journal of Medicinal Mushrooms. 2013;**15**(1):49-56

[45] Madan M, Thind KS. Physiology of Fungi. Ashish Publishing House - APH Publishing Corporation. BookVistas: New Delhi, India; 2000;**240**:59-50. (ISBN 10: 8170249414, ISBN 13: 9788170249412)

[46] Bettin F et al. Growth kinetics, production, and characterization of extracellular laccases from *Pleurotus sajor-caju* PS-2001. Process Biochemistry. 2011;**46**:758-764

[47] Zhang Z et al. Synergistic effect of thermostable beta-glucosidase TN0602 and cellulase on cellulose hydrolysis. 3 Biotech. 2017;**7**(1):54

[48] Zhang H, Xu Y, Yu S. Co-production of functional xylooligosaccharides and fermentable sugars from corncob with effective acetic acid prehydrolysis. Bioresource Technology. 2017;**234**:343-349

[49] Eichlerova I et al. Screening of *Pleurotus* ostreatus isolates for their ligninolytic properties during cultivation on natural substrates. Biodegradation. 2000;**11**(5):279-287

[50] Pandey AK et al. Production of ligninolytic enzymes by white rot fungi on lignocellulosic wastes using novel pretreatments. Cellular and Molecular Biology (Noisy-le-Grand, France). 2014;**60**(5):41-45

[51] Pant D, Adholeya A. Enhanced production of ligninolytic enzymes and decolorization of molasses distillery wastewater by fungi under solid state fermentation. Biodegradation. 2007;**18**(5):647-659

[52] Eichlerova I et al. Ligninolytic characteristics of *Pleurotus* ostreatus strain F6 and its monokaryotic protoplast derivative P19. Canadian Journal of Microbiology. 2000;**46**(12):1153-1158

[53] Malarczyk E, Jarosz-Wilkolazka A, Kochmanska-Rdest J. Effect of low doses of guaiacol and ethanol on enzymatic activity of fungal cultures. Nonlinearity in Biology Toxicology and Medicine. 2003;**1**(2):167-178

[54] Adebayo EA, Martinez-Carrera D. Oyster mushrooms (*Pleurotus*) are useful for utilizing lignocellulosic biomass. African Journal of Biotechnology. 2015;**14**(1):52-67

[55] Zuo S et al. Effect of Irpex lacteus, Pleurotus ostreatus and Pleurotus cystidiosus pretreatment of corn stover on its improvement of the in vitro rumen fermentation. Journal of the Science of Food and Agriculture. Aug 2018;98(11):4287-4295. DOI: 10.1002/jsfa.8951. Epub 2018 Mar 22

[56] Tao QQ et al. New sesquiterpenoids from the edible mushroom *Pleurotus cystidiosus* and their inhibitory activity against alpha-glucosidase and PTP1B. Fitoterapia. 2016;**111**:29-35

[57] Jayasuriya WJ et al. Hypoglycaemic activity of culinary *Pleurotus* ostreatus and *P. cystidiosus* mushrooms in healthy volunteers and type 2 diabetic patients on diet control and the possible mechanisms of action. Phytotherapy Research. 2015;**29**(2):303-309

[58] Lau CC, Abdullah N, Shuib AS. Novel angiotensin I-converting enzyme inhibitory peptides derived from an edible mushroom, *Pleurotus cystidiosus* O. K. Miller identified by LC-MS/MS. BMC Complementary and Alternative Medicine. 2013;**13**:313

Chapter 5

From a Sequential to a Continuous Approach for LVV-h7 Preparation during Enzymatic Proteolysis in a Microfluidic-Based Extraction Process

Kalim Belhacene, Ionela Ungureanu, Elena Grosu, Alexandra Blaga, Pascal Dhulster and Renato Froidevaux

Additional information is available at the end of the chapter

http://dx.doi.org/10.5772/intechopen.80228

Abstract

Intensification of process is increasingly interesting in the context of recovery of industrial wastes. Among these compounds, animal blood is underexploited although it is an important source of bioactive peptides. LVV-h7 (LVVYPWTQRF) is one of these bioactive peptides from bovine haemoglobin hydrolysate. Our innovative approach consists of a continuous process involving at microfluidic scale for enzymatic proteolysis of bovine haemoglobin by pepsin, selective extraction of LVV-h7 to an organic solvent during the enzymatic reaction, followed by a second extraction to an aqueous phase for organic solvent recycling. Thus, the obtainment of pure LVV-h7 peptide with an efficient methodology of extraction and solvent recycling was proved.

Keywords: haemoglobin, enzymatic hydrolysis, extraction, microfluidic, solvent recycling, bioactive peptides

1. Introduction

Intensification of process is considered as an indispensable part in the development of approach for obtaining product with high added value. The improvement of a process could concern the technology used, the safety or also the source of raw material employed. Among these sources, wastes of agricultural and food processing are considered as a cheap source of valuable components since the existent technologies allow the recovery of target compounds and their recycling [1]. The process was developed at industrial scale to valorise these kinds of products [2]. In this work, animal blood was studied as a source of bioactive peptides [3–6]. Some of these peptides revealed their potential as antimicrobial [7], opioid [8], antihypertensive [8] or antioxidative activities [9]. These peptides are generally obtained by enzymatic degradation of haemoglobin alpha and beta chains [3]. Among these peptides, one opioid peptide was studied, known as LVV-haemorphin-7. LVV-h7 corresponds to the amino acid sequence LVVYPWTQRF (Leu-Val-Val-Tyr-Pro-Trp-Thr-Gln-Arg-Phe), obtained from β chain of haemoglobin (β 31–40) by pepsin hydrolysis and well known as bioactive peptides involved in the treatment of humans diseases [8, 10–12].

LVV-h7 presents also interesting physical properties, especially hydrophobic character which makes it able to transfer from aqueous to organic media in liquid/liquid extraction process [13]. Previous studies were interested to extract this peptide during haemoglobin hydrolysis using water/butan-2-ol-octan-1-ol liquid/liquid biphasic system [14–16]. Even if this process has shown the ability to extract selectively LVV-h7, its implementation was very complex and laborious, due to the long time of the process carried out (more than 10 h) to obtain low extraction yield of peptide (about 5%), the control of the immobilised enzyme stability during the process to avoid its inhibition by solvents and a high quantity of solvent.

Keeping in mind the economic and environmental impacts of a process development which requires the use of organic solvent, the microfluidic domain could bring solutions to reduce these disadvantages. First, reduction of scale allows to improve the surface/volume ratio and thus the molecular transfer capacity while reducing the volume of solvent used and energy [17, 18]. Second, the implementation for continuous flow platforms brings advantages of liquid-liquid transfer favoured by laminar flow [19–21], in the case of non-miscible liquids, for molecular transfer. Several studies revealed the real advantage of using LLE in microfluidic system [19, 22–23].

Previous study realised on the enzymatic hydrolysis of haemoglobin by pepsin in a microfluidic reactor has shown the influence of the microfluidic scale to the kinetics acceleration of bioactive peptides appearance, such as LVV-h7, comparing to bench scale [24]. This study shows also the potential of combining enzymatic microreactor with liquid/liquid biphasic system for LVV-h7 extraction. Therefore, the approach that we propose is based on continuous aqueous haemoglobin hydrolysis by pepsin, LVV-h7 extraction towards an intermediate organic phase and LVV-h7 DES extraction in a receiving aqueous phase, allowing the solvent recycling. This integrated process with solvent recycling is presented in **Figure 1**. Before fully designing the entire microfluidic process, where all the reactions are carried out simultaneously, we separately investigated each reaction to notably determine their respective optimal ranges of conditions before combination as follows: (1) haemoglobin hydrolysis by pepsin occurs in a primary aqueous feed phase, generating the LVV-h7 in a very complex peptide mixture with more than one hundreds of peptides with different primary sequences; (2) the as-formed LVV-h7 is then selectively extracted into an organic solvent (octan-1-ol); (3) eventually, the LVV-h7 is des extracted in a second aqueous phase (called receiving aqueous phase), which also allows the octan-1-ol recycling in the extraction step. Our approach is not only focused on the compatibility issues of enzymatic catalysis and opioid peptide extraction, but also pays particular attention to integrating all the steps to minimise separation and recycling burdens, which can be detrimental for the overall economics and efficiency of the process. The methodology envisioned to move from a

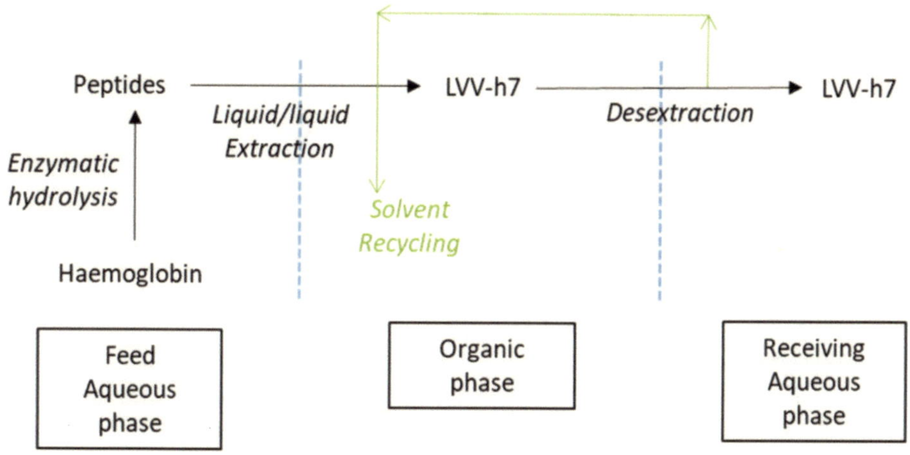

Figure 1. Simultaneous process applied to the haemoglobin hydrolysis by pepsin, the liquid/liquid extraction of LVV-h7 and the solvent recycling, to produce pure opioid peptide.

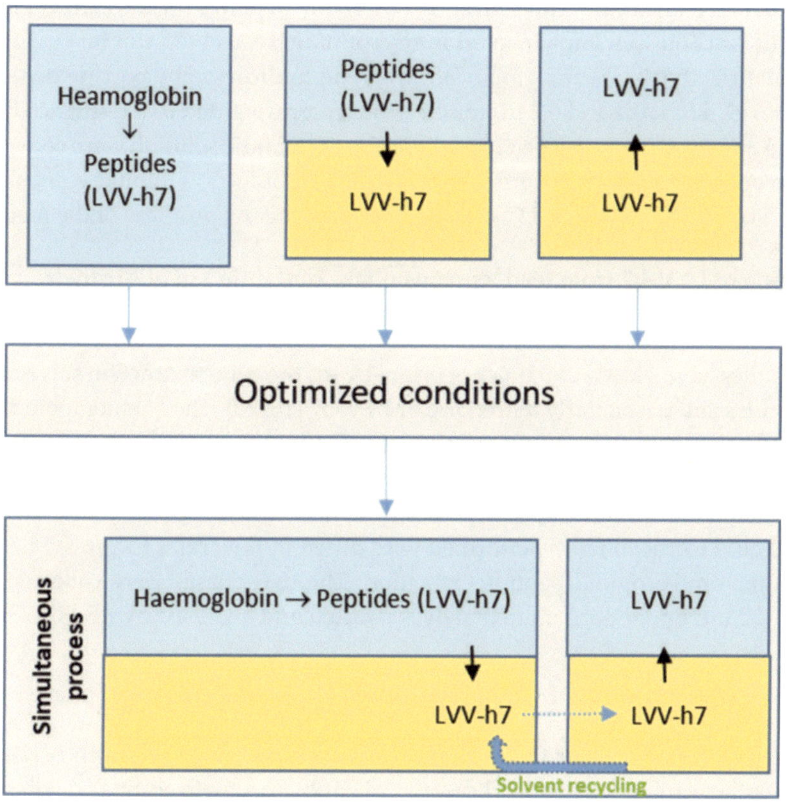

Figure 2. Methodology from a sequential approach towards an integrated continuous process.

sequential approach towards an integrated continuous process is schematised in **Figure 2**. Each step was optimised to obtain the best conditions concerning LVV-h7 concentration and purity, through residence time (i.e. flow rate) of both aqueous and organic phases inside the system.

After optimisation, enzymatic hydrolysis and extractions steps were combined for a continuous approach to obtain final aqueous phase containing the bioactive peptide. Quantification and purity of LVV-h7 were determined to evaluate the efficiency of integrated process.

2. Experimental procedure

2.1. Haemoglobin hydrolysis by pepsin

Bovine haemoglobin (64.5 kDa, Sigma Chemicals Co.) was hydrolysed by porcine pepsin (E.C. 3.4.23.1, 3.440 U mg^{-1}, 35 kDa, Sigma Chemicals Co.), protease from the family of aspartic acid proteases which preferentially catalyses the cleavage of peptide bonds at the carboxyl side of aromatic and hydrophobic amino acids. This proteolytic reaction leads to the appearance of product molecules called peptides. Haemoglobin was prepared under denaturing conditions at pH 3.0 by adding 2 M HCl for a 20 mg mL^{-1} final concentration. All aqueous solutions were prepared in 18.2 MΩ Milli-Q water (Millipore).

In order to obtain a maximum concentration of LVV-h7 peptides for the extraction step studies, kinetic of reaction was implemented in microfluidic system (75 μm inner diameter, 2 m length). Denatured bovine haemoglobin (1% w/v) was hydrolysed by porcine pepsin at different flow rates (9, 4.5, 1, 0.4 and 0.2 μL min^{-1}), which corresponded to a residence time of 15 s, 30 s, 2 min, 5 min and 10 min, respectively. Samples of peptidic solution were collected during the reaction and mixed with sodium hydroxide 1 M in order to inhibit the pepsin. Samples were conserved at 4°C for RP-HPLC analysis to determine the progress of the reaction.

2.2. Extraction of LVV-h7 from feed aqueous phase containing peptidic hydrolysate to octan-1-ol phase

Previous studies have shown the choice of octan-1-ol as the better extraction solvent for hydrophobic peptides and particularly in the case of LVV-h7 [16, 25]. The haemoglobin hydrolysate was pumped using peristaltic pump (Minipulls 3, Gilson Inc., Middleton, WI, USA) and sent in co-flow with octan-1-ol solution in the same capillary (75 μm inner diameter, 10 cm length) for liquid-liquid extraction (LLE) of peptides from haemoglobin phase to the organic phase. The liquid-liquid extraction was performed with different flow rates to appreciate the impact of contact time on the opioid peptide extraction. The two phases were collected out of the capillary in a 2 mL Eppendorf, immediately separated and analysed by RP-HPLC.

2.3. DES extraction of LVV-h7 from organic phase to receiving aqueous phase

Octan-1-ol phase obtained after the extraction procedure was pumped with peristaltic pump and fed in another capillary where acidic water (pH 3 obtained with acetic acid adjustment) was injected at different flow rates to perform a second liquid-liquid extraction from the octan-1-ol to the aqueous phase, called "DES extraction step". In these conditions, the acidic water extracts peptides and favours the extraction without mixing of the two phases [26]. Phases were collected out of the capillary in a 2 mL Eppendorf, immediately separated and analysed by RP-HPLC.

2.4. Coupling of extraction and DES extraction procedure

A coupling between both extractions optimised method was implemented, and its efficiency on LVV-h7 peptide extraction selectivity was measured. The entire system was represented on scheme **Figure 3** where the volume containing the peptidic hydrolysate and the octan-1-ol was pumped in the same capillary for extraction, followed by a pumping of octan-1-ol phase recovered in a second capillary, connected with acidic water for DES extraction of LVV-h7. For each part of the process, samples were collected for identification and quantification of species.

2.5. Reversed-phase-HPLC analysis

The liquid chromatographic system is consisted of a Waters 600E automated gradient controller pump module, a Waters Wisp 717 automatic sampling device and a Waters 996 photodiode array detector. Spectral and chromatographic data were stored in a NECImage 466 computer. Detection of the produced peptides was carried out at 215 nm by reverse phase HPLC (RP-HPLC) on a C4-column (Vydak 0.46 × 25 cm, 3 mm I.D.). The mobile phase was water/trifluoroacetic acid (100, 0.1, v/v) and acetonitrile/water/trifluoroacetic acid (60, 40, 0.1, by vol.) at 0.4 mL min^{-1} flow rate. All common chemicals and reagents were of analytical grade and were purchased from Sigma Chemicals Co. and Flandres Chimie. Identification and quantification of LVV-h7 were performed using peptide standard (purity: 91.63% M.W. 1308.56 Da) purchased from GeneCust Society (Luxembourg).

2.6. MALDI-TOF mass spectrometry analysis

The sample was loaded on a ground steel MALDI target (Bruker Daltonics, Bremen, Germany) following the dried droplet method. The MS (positive reflectron mode) and MS/MS (lift mode) measurements were performed in an automatic mode on an AUTOFLEXTM Speed

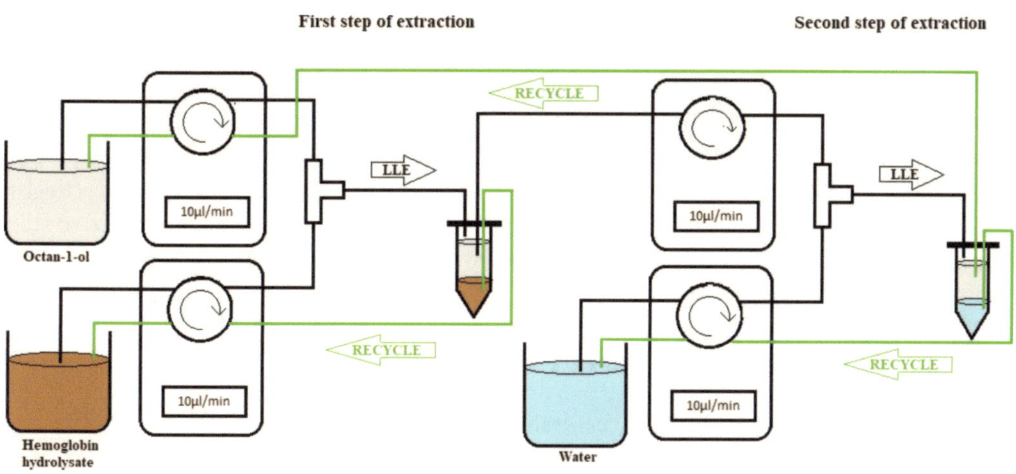

Figure 3. Implementation of microfluidic extraction and DES extraction steps of LVV-h7 from hydrolysate and recycling phases (black: employed method for extraction; green: recycling possibilities of solution).

TOF/TOF mass spectrometer (Bruker Daltonics) running FlexControlTM3.0 software (Bruker Daltonics). Peptide fragmentation was performed by automatic method of the manufacturer. MS and MS/MS spectra were processed using FlexanalysisTM3.3 and BioTools 3.4 software packages (Bruker Daltonics). Fragmentation pattern of peptides was deduced from matching of amino acid sequences (Uniprot accession numbers: P02070 & P01966) of each chain of bovine haemoglobin to the MS/MS spectra using BioTools 3.4.

3. Results and discussion

3.1. Haemoglobin hydrolysis by pepsin in feed aqueous phase

Haemoglobin 1% (w/v) was hydrolysed by porcine pepsin at room temperature, at pH 3 and at different residence time of substrate and enzyme in the microchannel. At the outlet of the capillary, samples were sent to an Eppendorf containing a disodium tetraborate buffer solution (0.32 M, pH 9.0), that caused enzyme denaturation and thus reaction stopping, and analysed by RP-HPLC.

Figure 4 shows the progressive decrease of alpha and beta chains of haemoglobin (between 40 and 60 min) to generate intermediate peptides (between 25 and 40 min) after 15 s of reaction. Next, after 30 s of reaction, there is an increase of peptide population between 25 and 30 min and an emergence of peptides between 15 and 25 min. For the last samples (5 and 10 min), there is a disappearance of peptides between 30 and 40 min to the profit of peptides between 5 and 25 min. This phenomenon was already described in the literature and was explained by an enzymatic hydrolysis of alpha and beta chains to produce a population of hydrophobic peptides with high molecular weight, observed between 30 and 35 min in **Figure 3** [3, 5]. Next, this population is hydrolysed to produce other peptides with intermediate molecular weight, observed between 20 and 30 min. Finally, these intermediate peptides are hydrolysed to generate small peptides with hydrophilic character (retention time between 5 and 20 min). This mechanism is called "zipper" mechanism, which is characterised by a denatured state of the initial haemoglobin structure. Moreover, this mechanism is more suitable for obtaining intermediate bioactive peptides, such as LVV-h7 (retention time of about 29 min in **Figure 4**), compared to a "one by one" mechanism where initial haemoglobin structure is in a native state [3, 5]. LVV-h7 was used as a standard to evaluate the efficiency of extraction process by octan-1-ol. Thus, for studying of the extraction step in microfluidic system, we decided to stop the reaction after 30 s of hydrolysis by sodium hydroxide 1 M in order to obtain the maximum quantity of LVV-h7 (**Figure 3c**).

3.2. Study of LVV-h7 extraction in octan-1-ol using microfluidic system

Haemoglobin hydrolysate whose concentration of LVV-h7 was the more important (1%, 30 s of hydrolysis with pepsin) was injected in co-flow (using a T connector) with octan-1-ol at different flow rates (5, 10, 20 and 50 μL min^{-1}). Samples were collected in Eppendorf, and octan-1-ol phase was analysed by RP-HPLC to highlight the extracted peptides. Results are shown in **Figure 5**.

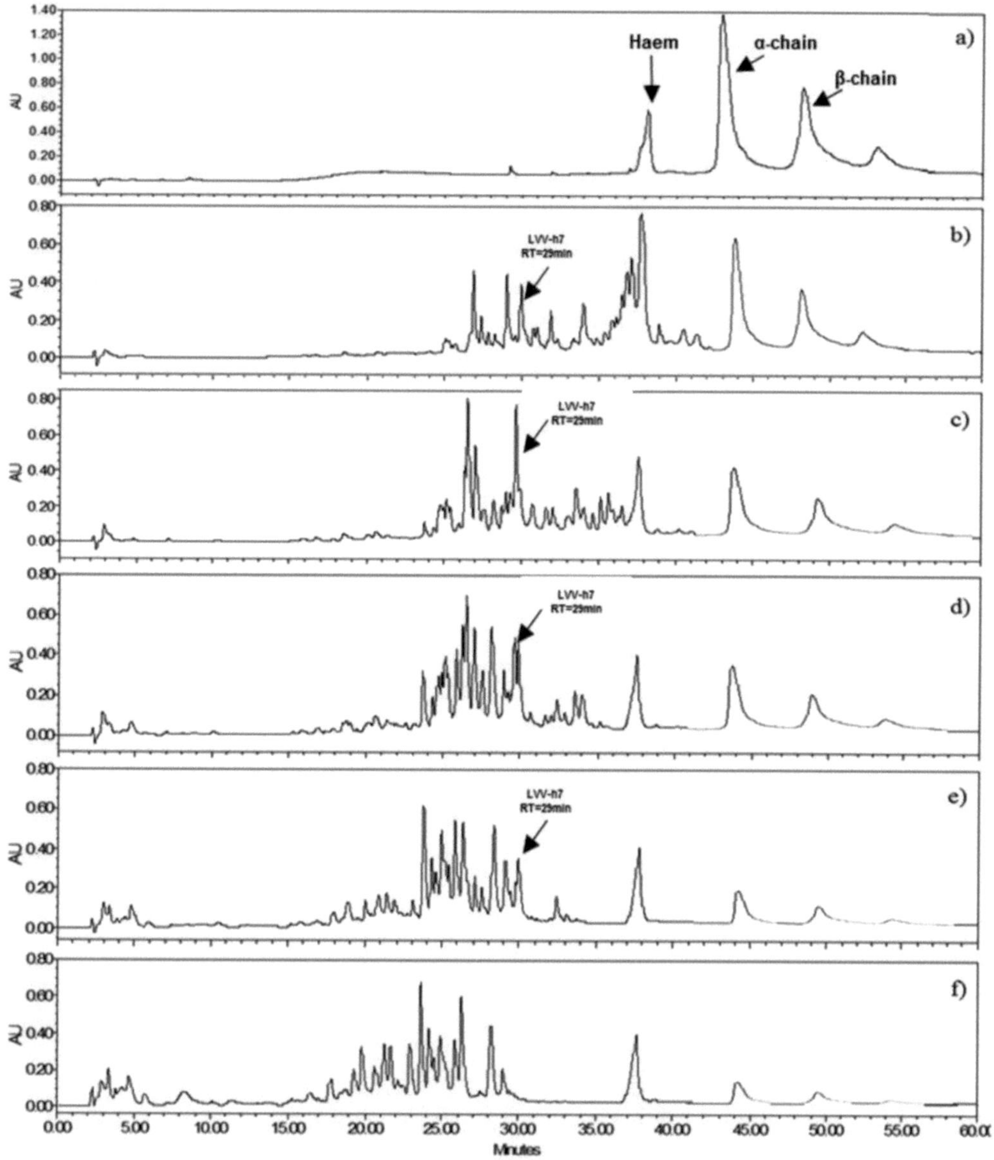

Figure 4. RP-HPLC chromatograms of haemoglobin 1% hydrolysed by porcine pepsin for different residence times of both haemoglobin and pepsin solutions. (a) Denatured haemoglobin without enzyme, (b) 15 s, (c) 30 s, (d) 2 min, (e) 5 min and (f) 10 min. Samples were analysed on C4 column.

The flow rates used for our analysis correspond to the additive flow rates of both solutions during the extraction step (i.e. initial flow rate multiplied by 2). Chromatograms show a good selectivity of the LLE method and octan-1-ol using microfluidic system. Indeed, we observe only three major peaks from the initial complex peptidic hydrolysate. Moreover, a predominance of LVV-h7 peptide is observed (RT of 29 min), which represents more than 55% of the peptides extracted by octan-1-ol. The selectivity of octan-1-ol for hydrophobic peptides such as LVV-h7 was previously confirmed by our team in batch and continuous reactors from enzymatic haemoglobin hydrolysates, but not in microfluidic system [14–16]. Two other peaks (RT of 27 and 31 min) are observed and supposed having a hydrophobic character.

The three peaks identified in the octan-1-ol phase were analysed by MALDI mass spectroscopy to appreciate their composition (**Figure 6**).

Concerning the fraction eluted at a 29 min (**Figure 6b**), result from mass spectroscopy shows unique peak at m/z = 1308.930 Da corresponding to molecular weight for LVV-h7 (with one hydrogen more from mass analysis). No other components were observed, which shows that LVV-h7 peptide is pure. The other fractions collected were also identified. At retention time

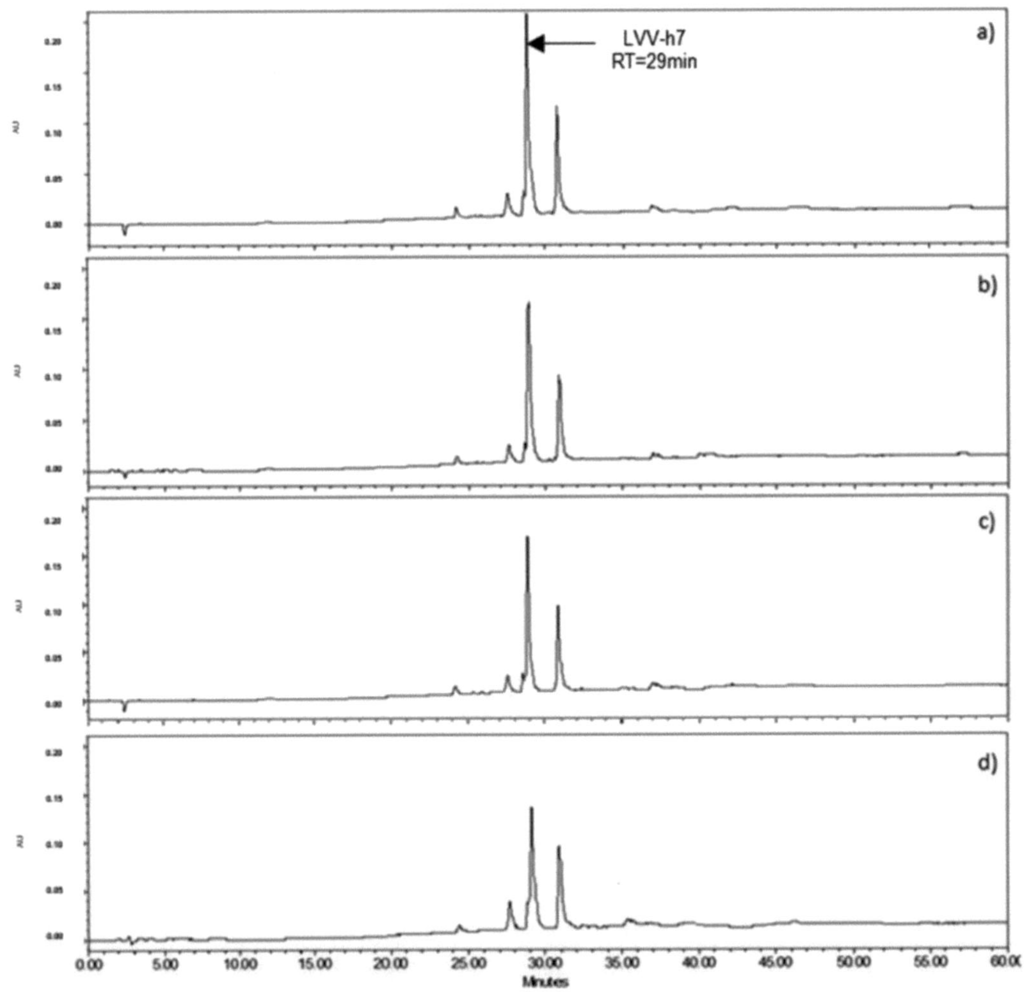

Figure 5. HPLC chromatograms of octan-1-ol phase after extraction applied on haemoglobin hydrolysate in microfluidic system at different flow rates ((a) 10 μL min^{-1}, (b) 20 μL min^{-1}, (c) 40 μL min^{-1} and (d) 100 μL min^{-1}).

of 27 min, another pure peptide with 1195.933 Da for molecular weight was identified. It corresponds to the VV-h7 peptide, also obtained by haemoglobin hydrolysis with pepsin and characterised as a hydrophobic bioactive peptide [7]. The fraction eluted at 31 min was composed of two major compounds (1422.138 and 1733.23 Da). These peptides correspond to other peptides without known biological activity.

Finally, using standard concentration curve (R^2 = 0.98) prepared with a pure standard of LVV-h7 (GeneCust, 91%), the quantity of pure LVV-h7 extracted from bovine haemoglobin hydrolysate was 6.12 ± 0.34 μg mL^{-1}. With an initial concentration of LVV-h7 calculated at 16.12 ± 0.85 μg mL^{-1} in the hydrolysate, performance of our system is around 38% of peptide extracted with only one cycle of extraction.

Table 1 resumes the total area of peptides extracted and particularly for LVV-h7 depending on the flow rate and thus on the residence time of hydrolysate in the microsystem. The area of peptides extracted from peptidic hydrolysate increases with the decreasing of flow rate used, confirming the influence of the time of contact between both phases in the capillary. A relative high time of contact between peptidic hydrolysate and octan-1-ol favours the diffusion of peptides to the solvent phase. However, even if the quantity of LVV-h7 obtained between 10

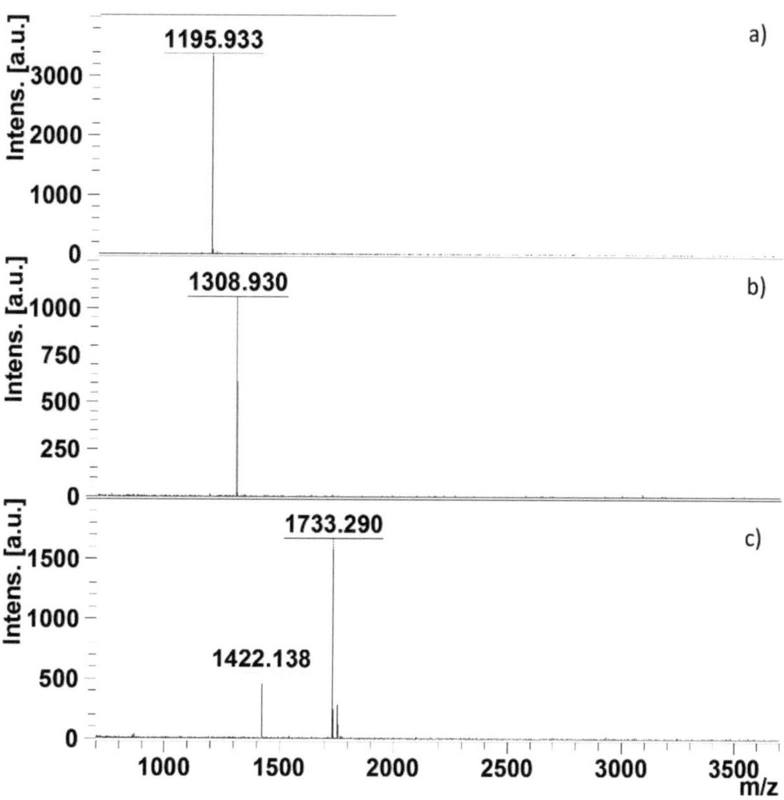

Figure 6. MALDI-TOF spectra obtained for each peak (elution time (a) 27 min, (b) 29 min and (c) 31 min) collected after octan-1-ol extraction and separated by RP-HPLC.

and 20 μL min^{-1} was relatively proportional (×1.5), this difference was less marked for flow rates used up to 20 μL min^{-1}.

3.3. Study of LVV-h7 DES extraction from octan-1-ol to receiving aqueous phase

Using the same process than in the first extraction (from peptidic hydrolysate to octan-1-ol), the octan-1-ol phase containing peptides was injected in co-flow with acidic aqueous phase in order to transfer these peptides from organic to receiving aqueous phase. Different flow rates were used to evaluate the impact of contact time between the two phases during the DES extraction process.

Chromatograms (**Figure 7**) show clearly a transfer of peptides from the organic phase to the aqueous phase by liquid-liquid DES extraction in the microfluidic conditions. The three major fractions corresponding to the peptides previously described and mainly present in the initial octan-1-ol phase were found in the receiving aqueous phase. Moreover, a small proportion of these peptides were still present in octan-1-ol after the DES extraction step. **Figure 8** illustrates the concentration of LVV-h7 in octan-1-ol and the receiving aqueous phase after the DES extraction for different flow rates.

The LVV-h7 concentration in the aqueous phase increases with the decrease of flow rate, indicating the influence of contact time between the two phases on the peptide diffusion. For 10 μL min^{-1}, more than 82% of the initial concentration of LVV-h7 phase was transferred in the aqueous phase. The calculation of LVV-h7 proportion in the aqueous phase compared to the total quantity of peptides extracted was 50 ± 1%. LVV-h7 fraction was analysed by MALDI mass spectroscopy to verify its purity (result not shown). Mass analysis reveals the purity of the LVV-h7 fraction to recover from organic phase to the receiving aqueous phase using microfluidic system. Both fractions (27 and 31 min) were also analysed by mass spectroscopy. It confirmed the same composition than obtained during the extraction step that is, a pure VV-h7 peptide for 27 min of elution time and other peptides for 31 min. Thus, the DES extraction step of peptides from octan-1-ol phase to an acidic aqueous phase using microfluidic system was validated with a good efficiency (more than 82% of transfer yield).

Flow rate (μL min^{-1})	Time of contact (s)	Total area of peptides (μV/*s)	LVV-h7 area (μV/*s)	Total % of LVV-h7[1]
10	2.65	10,793,556	6,292,478	58
20	1.32	6,976,911	4,064,156	58
40	0.66	6,316,966	3,599,507	57
100	0.26	5,921,488	3,181,856	54

[1]Total % LVV-h7 = (area of LVV-h7/total area) × 100)

Table 1. Proportion of LVV-h7 in total area of peptides after extraction process.

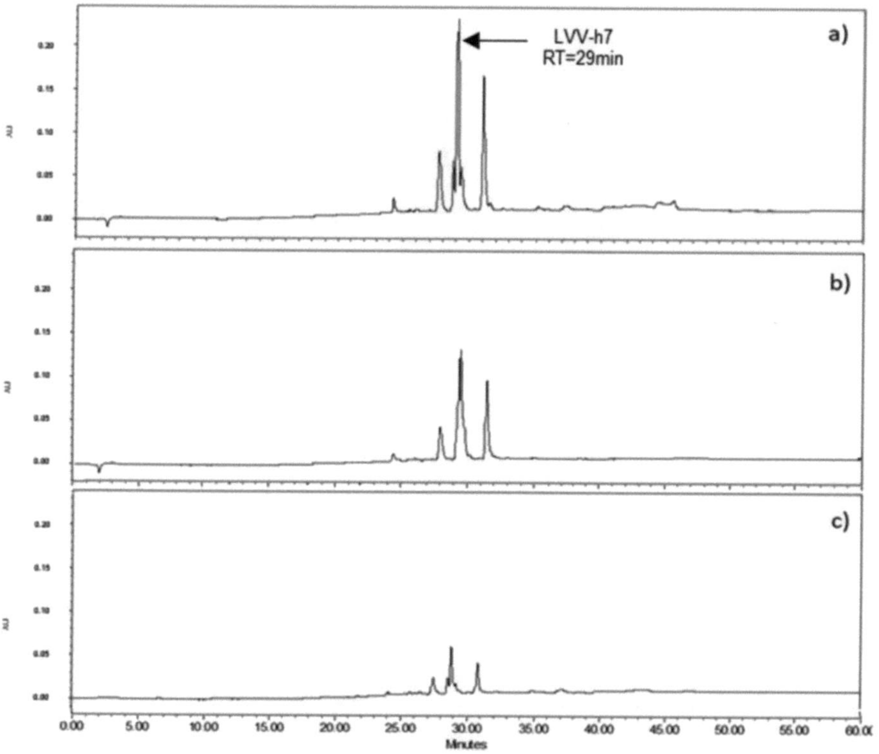

Figure 7. RP-HPLC chromatograms of octan-1-ol phase before DES extraction of peptides (a) water phase, (b) and octan-1-ol phase (c) obtained after DES extraction in microfluidic system for a flow rate of 20 μL min^{-1}.

3.4. Simultaneous haemoglobin hydrolysis by pepsin in feed aqueous phase, LVV-h7 extraction in octan-1-ol and DES extraction in receiving aqueous phase

In the optimal conditions previously determined, a continuous integrated process was tested. The microfluidic system, phases and matter flows are presented in **Figure 1**. The experiment was conducted following the same approach than those in **Figure 3** but with a feed aqueous phase formed by a haemoglobin solution, previously prepared under denaturing conditions at pH 3.0 (see part 3.1, haemoglobin final concentration of 1%, p/v), and a solution of pepsin with a E/S ratio of 1/11 (mol/mol). The introduction of these solutions in the microreactor (75 μm I.D. × 150 μm O.D.) was achieved by a syringe pump with a flow rate for both pepsin and haemoglobin solutions at about 4.5 μL min^{-1} for a capillary length of 2 m. The flow rates used correspond to the additive flow rates of both solutions, that is, initial flow rate multiplied by 2. The outlet fused silica capillary was in contact with a disodium tetraborate buffer solution (0.32 M, pH 9.0), thanks to a T-connection and a capillary (75 μm inner diameter, 10 cm length, flow rate of 5 μL min^{-1}), that caused enzyme denaturation and thus reaction stopping. Consequently to the enzymatic reaction, the resulting peptidic hydrolysate was sent in co-flow in a T-connector with

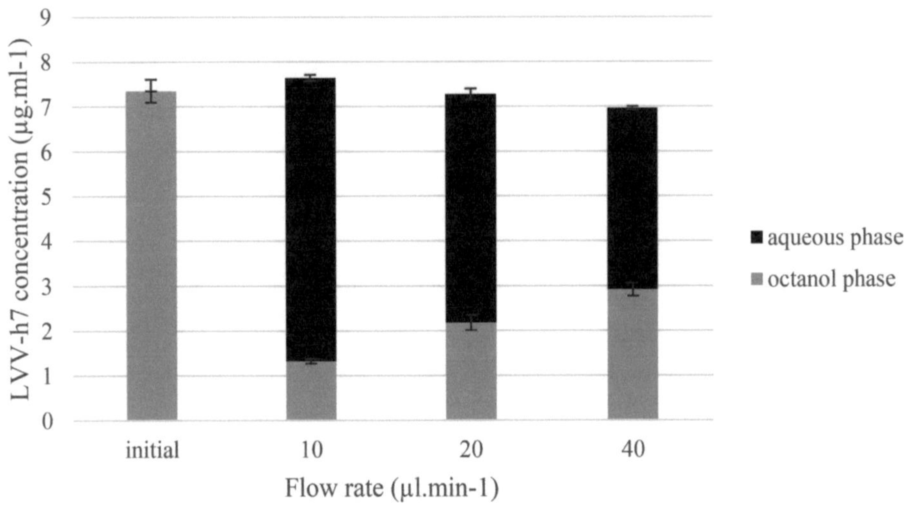

Figure 8. Proportion of LVV-h7 (µg mL^{-1}) in each phase after the DES extraction step in the microfluidic system for different flow rates.

octan-1-ol solution in a new capillary (75 µm inner diameter, 10 cm length) for LVV-h7 extraction at 10 µL min^{-1}. The two phases were collected out of the capillary in an Eppendorf. Octan-1-ol was pumped at 10 µL min^{-1} in a third capillary connected with acidic water for DES extraction of LVV-h7 thanks to a T-connector. Finally, the octan-1-ol phase remaining was pumped (10 µL min^{-1}) and re-injected in the initial octan-1-ol phase for a new cycle of extraction-DES extraction. To avoid the pumping of the bad phase, an offset of 10 min was done between each extraction step to have sufficient volume of phases. For each step of the continuous process, samples were collected for identification and quantification of species (**Figure 9** and **Table 2**).

First, the peptidic profile obtained during the enzymatic hydrolysis step confirms the results previously obtained by Elagli et al. [24] and the presence of LVV-h7 in the reaction medium (**Figure 9a**). Then, extraction of the opioid peptide to octan-1-ol phase is also observed in **Figure 9b**, showing the good selectivity of the organic phase for the same hydrophobic peptides detected before, whose LVV-h7 (**Figure 7**). Finally, we observe the haemorphin in the acidic receiving aqueous phase, confirming the DES extraction step efficiency. A proportion of the peptidic fractions is also found in the octan-1-ol phase after the DES extraction, translating an incomplete transfer in the receiving aqueous phase. However, results validate the process of simultaneous enzymatic hydrolysis of haemoglobin with LVV-h7 extraction and DES extraction at the microscale level.

The calculation of the different concentrations of LVV-h7 in each phase (**Table 2**) shows the transfer of 78% of the initial concentration of LVV-h7 from octan-1-ol to water. This result confirms the efficiency of the coupling approach of the extraction-DES extraction steps in microfluidic system after the enzymatic reaction. Moreover, after the DES extraction step, a low concentration of LVV-h7 remained in octan-1-ol phase (1.64 µg mL^{-1}). Thus, the reuse of octan-1-ol for a new cycle of extraction allows to obtain an efficient method of peptide recovery with a minimal of organic solvent quantity. An optimisation of the process, particularly the time of contact

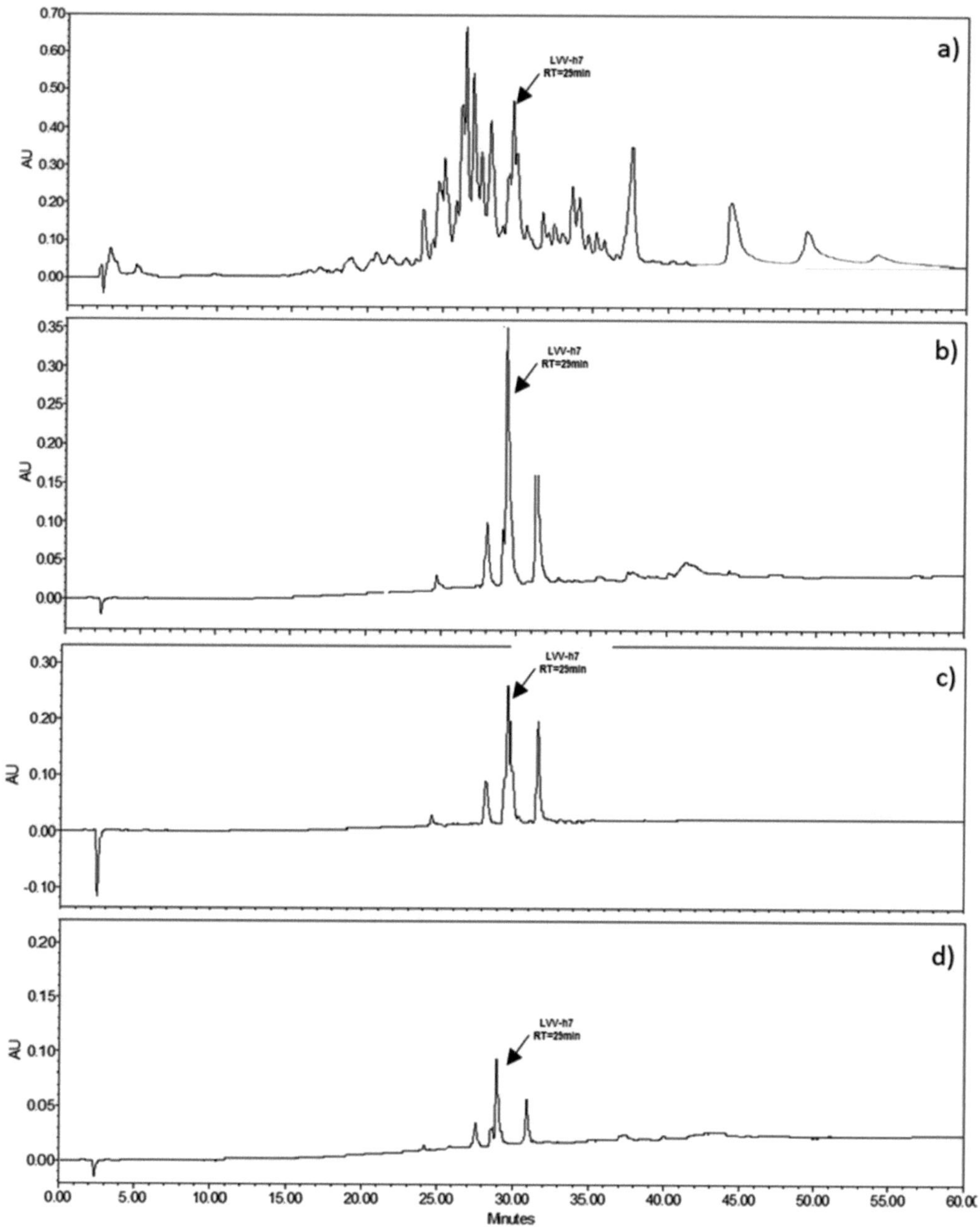

Figure 9. RP-HPLC chromatograms obtained from each step of the continuous process in the microfluidic system. (a) Peptidic hydrolysate after 30 s of haemoglobin hydrolysis by pepsin, (b) octan-1-ol phase after the extraction, (c) water phase after the DES extraction and (d) octan-1-ol phase after the DES extraction.

between feed aqueous/organic phases during the extraction and receiving aqueous/organic phases during the DES extraction could certainly improve the final concentration of LVV-h7 recovered and consequently decrease peptide concentration remained in the organic phase.

Phases collected	LVV-h7 area (μV/*s)	LVV-h7 concentration (μg mL^{-1})
Octan-1-ol (extraction step)	6,354,896	6.18
Octan-1-ol (DES extraction step)	1,687,646	1.64
Water (DES extraction step)	4,982,647	4.85

Table 2. Concentrations of LVV-h7 in each phase during the complete process.

4. Conclusion

In this work, we have first provided the conditions of the key parameters in microfluidic systems (residence times, flow rates, and concentrations) applied for a sequential process from liquid/liquid extraction of LVV-h7, present in a very complex peptidic hydrolysate, in octan-1-ol to its DES extraction in a second acidic aqueous phase. The optimised conditions have been then applied to an unprecedented integrated process in a specifically microfluidic approach. Therein, enzymatic hydrolysis of denatured haemoglobin by pepsin in a first microfluidic system was coupled with LVV-h7 extraction in octan-1-ol. This organic phase was then put in contact with a second aqueous phase for LVV-h7 DES extraction. The microfluidic scale allowed to increase the ratio surface/volume in order to favour the transfer of hydrophobic peptide to the organic solvent. A very good selectivity of extraction of the opioid peptide is obtained, from a very complex peptidic population generated during the enzymatic hydrolysis of haemoglobin by pepsin, with more than 38% of LVV-h7 initial concentration transferred to the organic phase. The DES extraction reveals also a very good transfer of LVV-h7 from octan-1-ol to the acidic aqueous phase, with more than 80%. Thus, the simultaneous process allowed to recover until more than 6 μg mL^{-1} of LVV-h7 with an excellent purity measured by mass spectroscopy with only one cycle of process (2.65 s of contact for each LLE).

The coupling of both extractions confirmed the feasibility of this process with a recycling of each phase to obtain a continuous process of extraction at microfluidic scale. An optimisation of the time of contact during the extraction is the key of peptide transfer between each phase and particularly for the recycling step. Currently, a study of a complete continuous process is in progress, where pepsin is immobilised in the microchannel and the outlet microcapillary is directly in contact with a solution of octan-1-ol, to avoid stopping the enzymatic reaction before extraction.

Acknowledgements

This work was supported by the CPER Alibiotech project, which is financed by European Union, French State and the French Region of Hauts-de-France.

Author details

Kalim Belhacene[1], Ionela Ungureanu[2], Elena Grosu[2], Alexandra Blaga[2], Pascal Dhulster[1] and Renato Froidevaux[1]*

*Address all correspondence to: renato.froidevaux@univ-lille1.fr

1 Université Lille, INRA, ISA, Université Artois, Université Littoral Côte d'Opale, EA7394-ICV-Institute Charles Viollette, Lille, France

2 Department of Organic, Biochemical and Food Engineering, Faculty of Chemical Engineering and Environmental Protection, Gheorghe Asachi Technical University of Iasi, Iasi, Romania

References

[1] Pfaltzgraff LA, De Bruyn M, Cooper EC, Budarin V, Clark JH. Food waste biomass: A resource for high value chemicals. Green Chemistry. 2013;**15**:307-314. DOI: 10.1039/C2GC36978H

[2] Narodoslawsky M. Chemical engineering in a sustainable economy. Chemical Engineering Research and Design. 2013;**91**:2021-2028. DOI: 10.1016/j.ece.2009.09.001

[3] Tavano OL. Protein hydrolysis using proteases: An important tool for food biotechnology. Journal of Molecular Catalysis B: Enzymatic. 2013;**90**:1-11. DOI: 10.1016/j.molcatb.2013.01.011

[4] In MJ, Chae HJ, Oh NS. Process development for heme-enriched peptide by enzymatic hydrolysis of haemoglobin. Bioresource Technology. 2002;**84**:63-68. DOI: 10.1016/S0960-8524(02)00009-3

[5] Nedjar-Arroume N, Dubois-Delval V, Yaba Adje E, Traisnel J, Krier F, Mary P, Kouach M, Briand G, Guillochon D. Bovine hemoglobin: An attractive source of antibacterial peptides. Peptides. 2008;**29**:969-977. DOI: 10.1016/j.peptides.2008.01.011

[6] Biesalki H, Dragsted LO, Elmadfa I, Grossklaus R, Müller M, Schrenk D, Walter P, Weber P. Bioactive compounds: Definition and assessment of activity. Nutrition. 2009;**25**:1202-1205. DOI: 10.1016/j.nut.2009.04.023

[7] Nedjar-Arroume N, Dubois-Delval V, Miloudi K, Daoud R, Krier F, Kouach M, Briand G, Guillochon D. Isolation and characterization of four antibacterial peptides from bovine haemoglobin. Peptides. 2006;**27**:2082-2086. DOI: 10.1016/j.peptides.2006.03.033

[8] Gomes I, Dale CS, Casten K, Geigner MA, Gozzo FC, Ferro ES, Heimann AS, Devi LA. Hemoglobin-derived peptides as novel type of bioactive signaling molecules. The AAPS Journal. 2010;**12**:658-668. DOI: 10.1208/s12248-010-9217-x

[9] Chang CY, Wu KC, Chiang SH. Antioxidant properties and protein compositions of porcine haemoglobin hydrolysates. Food Chemistry. 2010;**15**:1537-1543. DOI: 10.1016/j.foodchem.2005.12.019

[10] Blishchenko EY, Sazonova OV, Kalinina OA, Yatskin ON, Philippova MM, Surovoy AY, Karelin AA, Ivanov VT. Family of hemorphins: Co-relations between amino acid sequences and effects in cell cultures. Peptides. 2002;**23**:903-910. DOI: 10.1016/S0196-9781(02)00017-7

[11] John H, John S, Forssmann WG. Kinetic studies on aminopeptidase M-mediated degradation of human hemorphin LVV-h7 and its N-terminally truncated products. Journal of Peptide Science. 2008;**14**:797-803. DOI: 10.1002/psc.1002

[12] Froidevaux R, Lignot B, Nedjar-Arroume N, Guillochon D, Coddeville B, Ricart G. Kinetics of appearance of hemorphins from bovine hemoglobin peptic hydrolysates by a direct coupling of reversed-phase high-performance liquid chromatography and electrospray ionization mass spectrometry. Journal of Chromatography. A. 2000;**873**:185-194. DOI: 10.1016/S0021-9673(99)01353-9

[13] Zhao Q, Piot JM. Organic solvent extraction associated with HPLC in the preparation of hemorphins from bovine hemoglobin peptic hydrolysate. Preparative Biochemistry and Biotechnology. 1998;**28**:61-78. DOI: 10.1080/10826069808010127

[14] Froidevaux R, Vercaigne-Marko D, Kapel R, Lecouturier D, Chung S, Dhulster P, Guillochon D. Study of a continuous reactor for selective solvent extraction of haemorphins in the course of peptic haemoglobin hydrolysis. Journal of Chemical Technology and Biotechnology. 2006;**81**:1433-1440. DOI: 10.1002/jctb.1584

[15] Froidevaux R, Vanhoute M, Lecouturier D, Dhulster P, Guillochon D. Continuous preparation of two opioïd peptides and recycling of organic solvent using liquid/liquid extraction coupled with aluminium oxide column during haemoglobin hydrolysis by immobilized pepsin. Process Biochemistry. 2008;**43**:431-437. DOI: 10.1016/j.procbio.2008.01.006

[16] Vanhoute M, Froidevaux R, Vanvlassenbroeck A, Lecouturier D, Dhulster P, Guillochon D. Ion-pairing separation of bioactive peptides using an aqueous/octan-1-ol micro-extraction system from bovine haemoglobin complex hydrolysates. Journal of Chromatography B. 2009;**877**:1683-1688. DOI: 10.1016/j.jchromb.2009.04.011

[17] Aota A, Mawatari K, Kitamori T. Parallel multiphase microflows: Fundamental physics, stabilization methods and applications. Lab on a Chip. 2012;**9**:2470-2476. DOI: 10.1039/b904430m

[18] Zuloaga O, Olivares M, Navarro P, Vallejo A, Prieto A. Dispersive liquid–liquid microextraction: Trends in the analysis of biological samples. Bioanalysis. 2015;**12**:935-941. DOI: 10.4155/bio.15.141

[19] Jovanovi'c J, Rebrov EV, Nijhuis TA, Kreutzer MT, Hessel V, Schouten JC. Liquid–liquid flow in a capillary microreactor: Hydrodynamic flow patterns and extraction performance.

Industrial and Engineering Chemistry Research. 2012;**15**:1015-1026. DOI: 10.1021/ie200715m

[20] Jähnisch K, Hessel V, Löwe H, Baerns M. Chemistry in microstructured reactors. Angewandte Chemie, International Edition. 2004;**43**:406-446. DOI: 10.1002/anie.200300577

[21] Newman SG, Jensen KF. The role of flow in green chemistry and engineering. Green Chemistry. 2013;**15**:1456-1472. DOI: 10.1039/C3GC40374B

[22] Mu X, Liang Q, Hu P, Ren K, Wang Y, Luo G. Selectively modified microfluidic chip for solvent extraction of Radix Salvia Miltiorrhiza using three-phase laminar flow to provide double liquid–liquid interface area. Microfluidics and Nanofluidics. 2010;**9**:365-373. DOI: 10.1007/s10404-009-0554-y

[23] Zhao Y, Chen G, Yuan Q. Liquid–liquid two-phase flow patterns in a rectangular microchannel. AICHE Journal. 2006;**52**:4052-4060. DOI: 10.1002/aic.11029

[24] Elagli A, Laurette S, Treizebre A, Bocquet B, Froidevaux R. Diffusion based kinetic selectivity modulation of enzymatic proteolysis in a microfluidic reactor: Experimental analysis and stochastic modelling. RSC Advances. 2014;**4**:3873-3882. DOI: 10.1039/C3RA46005C

[25] Ontiveros JF, Froidevaux R, Dhulster P, Salager JL, Pierlot C. Haem extraction from peptidic hydrolysates of bovine haemoglobin using temperature sensitive C10E4/O/W micro emulsion system. Colloids and Surfaces A: Physicochemical and Engineering Aspects. 2014;**454**:135-143. DOI: 10.1016/j.colsurfa.2014.04.002

[26] Mason LR, Ciceri D, Harvie DJE, Perera JM, Stevens GW. Modelling of interfacial mass transfer in microfluidic solvent extraction: Part I. Heterogenous transport. Microfluids and Nanofluids. 2013;**14**:197-212. DOI: 10.1007/s10404-012-1038-z